T0195522

Warfighting and Logistic Support of Joint Forces from the Joint Sea Base

Robert W. Button • John Gordon IV • Jessie Riposo
Irv Blickstein • Peter A. Wilson

Prepared for the United States Navy
Approved for public release; distribution unlimited

 NATIONAL DEFENSE RESEARCH INSTITUTE

The research described in this report was prepared for the United States Navy. The research was conducted in the RAND National Defense Research Institute, a federally funded research and development center sponsored by the Office of the Secretary of Defense, the Joint Staff, the Unified Combatant Commands, the Department of the Navy, the Marine Corps, the defense agencies, and the defense Intelligence Community under Contract W74V8H-06-C-0002.

Library of Congress Cataloging-in-Publication Data is available for this publication.

ISBN 978-0-8330-4195-1

The RAND Corporation is a nonprofit research organization providing objective analysis and effective solutions that address the challenges facing the public and private sectors around the world. RAND's publications do not necessarily reflect the opinions of its research clients and sponsors.

RAND® is a registered trademark.

Cover Design by Rod Sato

© Copyright 2007 RAND Corporation

All rights reserved. No part of this book may be reproduced in any form by any electronic or mechanical means (including photocopying, recording, or information storage and retrieval) without permission in writing from RAND.

Published 2007 by the RAND Corporation
1776 Main Street, P.O. Box 2138, Santa Monica, CA 90407-2138
1200 South Hayes Street, Arlington, VA 22202-5050
4570 Fifth Avenue, Suite 600, Pittsburgh, PA 15213-2665
RAND URL: http://www.rand.org/
To order RAND documents or to obtain additional information, contact Distribution Services: Telephone: (310) 451-7002;
Fax: (310) 451-6915; Email: order@rand.org

Preface

Sea Basing is a fundamental concept to the Navy's operational vision for the 21st century. Navy–Marine Corps concepts for Sea Basing would enable joint force commanders to accelerate deployment and employment of naval power-projection capabilities. The overall intent of Sea Basing is to use the flexibility and protection provided by the sea base while minimizing the presence of forces ashore. The Assessment Division of the Office of the Chief of Naval Operations (N81) of the U.S. Navy asked the RAND Corporation to examine how still-evolving Navy–Marine Corps concepts for Sea Basing could be applied to joint operations beyond the Department of the Navy. N81 particularly desired insights on the use of Sea Basing to support Army operations.

This monograph presents the results of research performed by the RAND National Defense Research Institute for N81. It should be of interest to the Department of the Navy, the Department of the Army, the Office of the Secretary of Defense (OSD), and Congress.

This research was conducted within the Acquisition and Technology Policy Center of NDRI, a federally funded research and development center sponsored by the Office of the Secretary of Defense, the Joint Staff, the Unified Combatant Commands, the Department of the Navy, the Marine Corps, the defense agencies, and the defense Intelligence Community.

For more information on RAND's Acquisition and Technology Policy Center, contact the Director, Philip Antón. He can be reached by email at atpc-director@rand.org; by phone at 310-393-0411, extension

7798; or by mail at the RAND Corporation, 1776 Main Street, Santa Monica, California 90407-2138. More information about RAND is available at www.rand.org/.

Contents

Figures

Tables

Summary

Sea Basing, a fundamental concept in Sea Power 21, the Navy's operational vision for the 21st century, is designed to help joint force commanders accelerate deployment and employment of naval power and to enhance seaborne positioning of joint assets. It will do so by minimizing the need to build up a logistics stockpile ashore, reducing the operational demand for sealift and airlift assets, and permitting forward positioning of joint forces for immediate employment.

The cornerstone of sea-based logistics on the brigade scale is the Maritime Pre-positioning Force and its future version, the MPF(F). The Maritime Pre-positioning Force currently consists of three forward-deployed squadrons of maritime pre-positioning ships, each with five or six vessels with weapons, supplies, and equipment sufficient to support a force about the size of a Marine Expeditionary Brigade for up to 30 days. The MPF(F) will be composed of multiple ship types designed to support a Marine Expeditionary Brigade and provide functions not currently provided by the MPF, such as at-sea arrival, assembly, sustainment, reconstitution, and redeployment of Expeditionary Forces, as well as Expeditionary Strike Group interoperability. Current plans call for an MPF(F) squadron comprising three large-deck amphibious ships, three Mobile Landing Platform transport ships,[1] and eight cargo ships.

The Assessment Division of the Office of the Chief of Naval Operations (OPNAV N81) asked the RAND Corporation's National

---

[1] The Mobile Landing Platform is a new-design ship that will carry Landing Craft Air Cushion (LCAC) connectors for the MPF(F). The LCAC is similar to a large hovercraft.

Defense Research Institute to examine how the still-evolving concepts for sea basing could be applied to joint operations. The Navy is particularly interested in how the sea base could support Army operations while supporting Marine Corps operations. This monograph provides a high-level analysis of the sea base, its use in operations related to the Marine Corps, and the viability of Army operations using the sea base under varying conditions.[2] This effort is not a definitive logistics-based study. Rather, it is conceptual in nature and uses a broad-brush model to define throughput capacity (and overcapacity, as discussed below).

The Army has historically deployed its forces for overseas conflicts by sea, a concept it has again recently emphasized. Although the Army emphasizes deploying its forces directly into an area of operations, rather than through at-sea assets, such as the MPF(F), the capability to perform at-sea transfer of Army forces could greatly benefit the joint force, particularly by providing a means to rapidly introduce Army forces where a usable port is not available.

Analysis and Scenarios

We examined three operational scenarios, in addition to support of a Marine Expeditionary Brigade (MEB) alone, that explore potential joint operations using the sea base to (1) support an Army light or airborne brigade that arrives 50 nautical miles (NM) inland in an area of operations, (2) support an Army medium (Stryker) or heavy brigade that arrives through a seaport of debarkation, and (3) move ashore an Army medium or heavy brigade that deploys through the sea base to the area of operations. In our analysis, we always assumed that the MPF(F) would support the MEB as its first priority. Once that mission was accomplished, any remaining capacity was identified as potentially available to support other joint forces—specifically, Army brigades of various types. Our analysis concluded that, in many circumstances, brigade-level Army and Marine Corps ground elements can be sus-

---

[2] In operations involving both the Marine Corps and the Army, the joint force commander will determine how and when they will use a sea base.

tained simultaneously using the throughput capacities of planned MPF(F) components.

The Seabasing Joint Integrating Concept, in its assessment of seabasing risks, states, "Adverse weather conditions and sea state impact sea-based operations can affect the rapid build-up of combat power and timely sustainment of employed forces"[3] Issues of sustainment under unfavorable conditions, such as in high sea states with degraded ship-to-ship movement, can be addressed, in part, using the metric of *relative sustainment capacity*, defined as the ratio of maximum sustainment throughput capacity (in short tons per day) to sustainment requirement (also in short tons per day).[4] Overcapacity exists under favorable conditions when this ratio exceeds 100 percent. Overcapacity is needed to ensure adequate capacity under unfavorable conditions. Overcapacity can also release some sea base assets (notably, MV-22 aircraft) for support to ground forces under favorable conditions.

Our analysis began with the collection of data from the Army, Navy, and Marine Corps. Related studies were also collected and examined. We developed three illustrative scenarios judged most likely to represent logistic support to Marine Corps and Army ground elements. We then developed a simulation, the Joint Sea Based Logistics Model (described in Appendix E), to quantify the capabilities of the sea base in these three scenarios. This simulation was used for hundreds of combinations of distances, ground elements to be sustained, levels of combat, possibilities for reducing sustainment demand, and various ship-to-shore connector assets. Our insights and recommendations derive both from simulation results and from an improved understanding of sea-based logistic support. They led to the following distinct approaches to increasing sustainment capacity:

---

[3] Department of Defense, *Seabasing Joint Integrating Concept*, Version 1.0, Washington, D.C., August 2005, p 12.

[4] For presentation purposes, our analysis consolidates all sustainment and lift requirements using the simple metric of tons per day. The underlying analysis considers classes of sustainment.

- Reducing distances from the sea base to supported ground elements or seaports of debarkation. Reducing sustainment distances from the planned distance of 110 NM is the most effective means of increasing sustainment capacity. Threat conditions can limit this option, necessitating others.
- Adding LCAC surface connectors to CH-53 and MV-22 aircraft in sustainment. The addition of LCACs could more than double sustainment throughput.[5]
- Increasing the ratio of CH-53K to MV-22 aircraft. The benefits of increasing the ratio of CH-53K to MV-22 aircraft can be similar to those from adding LCACs as sustainment assets.
- Reducing sustainment requirements. Reducing demand for external sustainment, such as that realized by eliminating ground elements' demand for bulk water, can significantly improve the ability to sustain ground elements.

We identified the following approaches to reducing Army ground element movement time from the sea base ashore:

- Increasing the ratio of CH-53K to MV-22 aircraft. A modest reduction in movement time for Army forces can be achieved by increasing the ratio of CH-53K to MV-22 aircraft. Put another way, such a change would, as described above, enhance sustainment performance significantly without increasing movement time.
- Adding Joint High-Speed Vessels to augment LCACs as surface connectors. Adding a single Joint High-Speed Vessel to augment LCACs roughly doubles surface connector throughput capacity and halves the movement time of Army brigade combat teams.

---

[5] Maintenance requirements limit LCACs to not more than 16 hours of operation per day. Crew fatigue can further limit LCACs to 12 hours or less of operation per day. Sixteen-hour days are used as a baseline for LCAC operations in the main body of this monograph; 12-hour days are considered as an excursion in Appendix E.

Sustainment Findings

Our analysis indicates that a Sea Base Maneuver Element, that portion of a Marine Expeditionary Brigade projected ashore for operations, can be sustained with some difficulty at a range of up to 110 NM from the sea base, using only CH-53K and MV-22 aircraft. Simultaneously sustaining both a Shore Based Maneuver Element and an Army airborne brigade using only these aircraft would require reducing significantly the distance from the sea base to these forces.

Using LCACs to augment sea base aircraft in sustainment has substantial benefits, particularly when LCACs contribute to both Marine Corps and Army ground element sustainment. When LCACs can contribute only to Marine Expeditionary Brigade sustainment, the limitations of airborne sustainment to Army ground elements determine the feasibility of joint sustainment. The use of a mix of sea base aircraft more rich in CH-53K aircraft than currently planned could enable joint sustainment at greater distances.

Reducing sustainment demand (by, for example, eliminating demand for bulk water from the sea base) is particularly helpful when sustainment capacity is marginal.

Movement Findings

An Army Stryker or heavy brigade can be transloaded at sea[6] and moved ashore from the sea base in three to six days (depending on the distance off shore), using MPF(F) assets also sustaining a MEB. The ability to move an Army brigade ashore in a few days represents a new capability for the Army.

If a single Joint High-Speed Vessel can augment the LCACs, it will roughly halve the time required to transport an Army brigade ashore. This finding reflects the observation that, when operable, the throughput capacity of a single Joint High-Speed Vessel about matches

---

[6] *Transloading* entails ship-to-ship movement by ramp. Transloading operations are illustrated in Figures B.3 and B.4.

the combined throughput of MLP LCACs. There are, however, issues of Joint High-Speed Vessel operability in this role in even moderate sea states, as well as the need for a small port where the Joint High-Speed Vessel can offload.

Other Findings

- The CH-53K is better suited than the MV-22 for sustainment; with external loads the MV-22 loses its speed advantage on ingress and the CH-53K carries at least twice the load of the MV-22. CH-53K helicopters are especially valuable under conditions of heavy sustainment demand or long sustainment distances.
- The Sea Basing concept is not consistent with, and in some sense conflicts with, the Army's desire to deploy directly to a port via High-Speed Ships. The Army has not developed doctrine and has not funded systems for operating with sea bases. However, our analysis illustrates that, once ashore, an Army brigade could, in many situations, be sustained by a sea base if (1) it moves away from its port of debarkation or (2) enemy action causes that port to become unavailable for sustainment.
- To capitalize on the potential of the sea base, Army shipping should be configured for "selective offload" rather than "dense pack." The interface between Army pre-positioning ships and the MLP is a potential bottleneck in moving Army forces. To avoid such bottlenecks, a built-in loading system should be considered for the MLP. Integrating such a loading system into the MLP might be less expensive in net than integrating it into Army and Navy pre-positioning ships and might also hasten joint interoperability.
- MPF(F) ships can provide deck space for a limited number of Army helicopters on a temporary basis (1–2 deck spots per "big deck") without significant loss of throughput capacity. However, there is not sufficient space on the MPF(F) to base significant numbers of Army aircraft as long as large numbers of Marine Corps MV-22 and CH-53K aircraft are based on these ships. Space for Army

aircraft could be created temporarily by moving MV-22 aircraft ashore, but several problems would remain, including rotor issues (braking and folding), corrosion, and maintenance.

Key Assumptions

To conduct the analysis, a number of assumptions were made. They included the following:

- Army unit equipment and supplies arrive at the sea base via Army shipping. Therefore, the Army units would not consume the MEB's supplies that are on the MPF(F) ships.
- Army ships arrive at the sea base "combat loaded" for selective offload, as opposed to "dense packed." Combat loaded ships are filled to roughly 60–70 percent of capacity in order to provide room to move vehicles and equipment below decks so that a specific item can be offloaded when needed. On the other hand, "dense packed" ships are loaded in a manner to maximize their carrying capacity. In that case, the ship can unload cargo only in the reverse order from how it was placed in the ship (i.e., the first piece of cargo loaded deep inside the ship will be the last item that can be removed).
- The connectors (e.g., ramps) between the Army's ships and the Mobile Landing Platform vessels will permit the movement of Army vehicles onto the MLP and its LCACs. Additionally, we assume that Army vehicle drivers would be properly trained to move their vehicles on board ships, including onto connecting ramps between ships.
- When LCACs are used to move Army and Marine Corps supplies ashore, sufficient trucks are available to move those supplies inland to where they would be consumed, and those trucks are adequately protected. It should be noted that an examination of the required number of trucks was not part of this analysis for the Navy. This issue, however, clearly merits more detailed analysis.

Acknowledgments

This study benefited from discussions with and data provided by LCDR Jeffrey Sinclair (OPNAV N81MF), CAPT Robert Winsor (OPNAV N81M), LCDR Frank Futcher (OPNAV N42), John Kaskin (OPNAV N42), CAPT James Stewart (OPNAV N42), Al Sawyers (U.S. Marine Corps MCCDC), LTC James R. Young (U.S. Army Combined Arms Support Command), Ed Horres (U.S. Army Training and Doctrine Command), and Michael W. Smith (Center for Naval Analyses).

We thank Clifford Grammich for his skillful support in the preparation of graphics for this monograph and for improving its readability. Finally, we thank ADM Don Pilling, USN (Ret.), and John Friel for their thoughtful reviews of this study, which benefited from their insights.

Acronyms

ABN	airborne
ADC(X)	Auxiliary Dry Cargo Carrier
AoA	Analysis of Alternatives
APOD	aerial port of debarkation
BCT	Brigade Combat Team
BLT	Brigade Landing Team
C2	Command and control
CDD	Capabilities Development Document
CLF	Combat Logistics Force
CNA	Center for Naval Analyses
CNO	Chief of Naval Operations
CONOP	Concept of Operations
CONUS	Continental United States
CSG	Carrier Strike Group
DOS	days of supply
DS	dry stores
EFSS	Expeditionary Fire Support System

EFV	Expeditionary Fighting Vehicle
ESG	Expeditionary Strike Group
FBE	Forward Base Echelon
FCS	Future Combat System
HBCT	heavy Brigade Combat Team
HMMWV	High-Mobility Multipurpose Wheeled Vehicle
H2O	water
HSC	high speed surface connector
HSS	High-Speed Ship
IBCT	Infantry Brigade Combat Team
ISO	International Standards Organization
ITV	Internally Transported Vehicle
JHSV	Joint High-Speed Vessel
JLOTS	Joint Logistics Over the Shore
JSF	Joint Strike Fighter
JSLM	Joint Seabasing Logistics Model
JTRS	Joint Tactical Radio Set
LAV	Light Armored Vehicle
LCAC	Landing Craft Air Cushion
LCU	Landing Craft Utility
LHA	Amphibious Assault Ship, general purpose
LHA(R)	LHA(Replacement)
LHD	Amphibious Assault Ship, multipurpose

LMSR	Large Medium-Speed Roll-on/Roll-off
LVS	Logistics Vehicle System
MAGTF	Marine Air-Ground Task Force
MCCDC	Marine Corps Combat Development Command
MEB	Marine Expeditionary Brigade
MEU	Marine Expeditionary Unit
MLP	Mobile Landing Platform
MPF	Maritime Pre-positioning Force
MPF(F)	Maritime Pre-positioning Force (Future)
MPG	Maritime Pre-positioning Group
MPSRON	Maritime Pre-positioning Ship Squadron
MTVR	Medium Tactical Vehicle Replacement
MV	motor vessel
NDIA	National Defense Industrial Association
NM	nautical mile
NRAC	Naval Research Advisory Committee
OPNAV	Office of the Chief of Naval Operations
POL	Petroleum, Oil, and Lubricants
PSYOPS	psychological operations
Recon	reconnaissance
RSO&I	reception, staging, onward movement, and integration
RSTA	reconnaissance, surveillance, and target acquisition

SBCT	Stryker Brigade Combat Team
SBE	Sea Base Echelon
SBME	Sea Base Maneuver Element
SBSE	Sea Base Support Element
SPOD	seaport of debarkation
ST	short ton
STOM	Ship-to-Objective Maneuver
T-AKE	dry cargo/ammunition ship
TSV	Theater Support Vessel
TUAV	tactical unmanned aerial vehicle
UAV	unmanned aerial vehicle
VERTREP	vertical replenishment
VTOL	Vertical Takeoff and Landing Aircraft
USAWC	U.S. Army War College
USMC	United States Marine Corps
USN	United States Navy
USNS	U.S. Naval Ship

CHAPTER ONE

Introduction and Objectives

Introduction

Sea Basing is a fundamental concept in Sea Power 21, the Navy's operational vision for the 21st century. The overall intent of Sea Basing is to make use of the flexibility and protection provided by the sea base while minimizing the presence of forces ashore. Sea Basing will enable joint force commanders to accelerate deployment and employment of naval power-projection capabilities and will enhance seaborne positioning of joint assets. It will also minimize the need to build up a logistics stockpile ashore, reduce the operational demand for sealift and airlift assets, and permit forward positioning of joint forces for immediate employment.[1]

Study Objectives

The Assessment Division of the Office of the Chief of Naval Operations (OPNAV N81) asked the RAND Corporation's National Defense Research Institute to examine how the still-evolving Navy–Marine Corps concepts for sea basing could be applied to joint opera-

---

[1] Formally, "the sea base of the future will be an inherently maneuverable, scalable aggregation of distributed, networked platforms that enable the global power projection of offensive and defensive forces from the sea, and includes the ability to assemble, equip, project, support, and sustain those forces without reliance on land bases within the Joint Operations Area" (Department of Defense, *Sea Basing Joint Integrating Concept*, Version 1.0, Washington, D.C., August 2005, p. 18).

tions beyond the Department of the Navy. The Navy was particularly interested in gaining insights on how the sea base could support Army operations.

Study Approach

The study began with the collection of data from the Army, Navy, and Marine Corps. Related studies were also assembled and examined. We developed three illustrative scenarios judged most likely to represent logistic support to Marine Corps and Army ground elements. We then developed a simulation, the Joint Sea Based Logistics Model (described in Appendix E), to quantify the capabilities of the sea base in these three scenarios. This simulation was used for hundreds of combinations of distances, ground elements to be sustained, levels of combat, possibilities for reducing sustainment demand, and various ship-to-shore connector assets. Our insights and recommendations derive both from simulation results and from an improved understanding of sea-based logistic support.

Organization of This Report

Chapter Two describes Army and Marine Corps operational concepts related to sea basing. It then introduces and discusses three operational scenarios intended to represent most likely cases for Army (airborne, Stryker, and heavy) brigade interaction with a sea base. Chapter Three presents a quantitative analysis of these three scenarios to determine factors in sea base performance and the value of related assets from outside the sea base—specifically, the Joint High-Speed Vessel (JHSV). Chapter Four draws together conclusions from the study.

Appendix A provides analytic results for additional cases and amplifies some findings in the main body of this monograph. Appendix B describes the Maritime Pre-positioning Force (Future) (MPF(F)) vessels in this analysis. Appendix C describes Army and Marine Corps ground elements in this study. Appendix D describes sustainment requirements for the ground elements described in Appendix C. Appendix E describes the primary analytic tool for this study, the Joint Seabasing Logistics Model (JSLM).

CHAPTER TWO

Operational Concepts and Scenarios

Background

Sea Basing is not an entirely new concept; Carrier Strike Groups (CSGs) and Expeditionary Strike Groups (ESGs) are sea bases. Indeed, during World War II the United States conducted several large-scale operations in which all the fire and logistic support was provided from offshore Navy ships. Scalability is a critical new element of the Sea Basing construct: whereas an ESG can support a Marine Expeditionary Unit (MEU) from the sea, future sea bases are expected to support one or more Marine Corps or Army brigades. Logistic sustainment concepts and their implementation are therefore key challenges in Sea Basing. The cornerstone of sea-based logistics on the brigade scale is the Maritime Pre-positioning Force (MPF) and its future version, the MPF(F).

MPF

The MPF currently consists of 16 ships organized into three forward-deployed Maritime Pre-positioning Ship Squadrons (MPSRONs). Each MPSRON consists of five or six ships loaded with pre-positioned weapons, supplies, and equipment sufficient to support a Marine Expeditionary Brigade (MEB)-sized Marine Air-Ground Task Force (MAGTF) (approximately 17,000 Marines) for up to 30 days.

Current MPF doctrine is to pre-position caches of supplies and oversized equipment at strategic locations. Forces are assembled and integrated through a cycle of reception, staging, onward movement, and integration (RSO&I). In the reception phase, a deploying joint force is airlifted into theater and received at an aerial port of debarkation (APOD). Simultaneously, MPF ships loaded with the deploying force's equipment arrive at a seaport of debarkation (SPOD). In the staging phase, deploying forces join with their equipment in marshalling areas near the SPOD. Onward movement is accomplished when the force departs the staging areas and moves to its assigned area of operations. Finally, integration occurs when the combat force commander places the force in his order of battle. Sustainment of the deployed force begins once it is received and transported to its staging areas and continues until the campaign is completed.

Operation Desert Storm fully demonstrated the MPF concept; MPF operations provided the first self-sustaining, operationally capable force in northern Saudi Arabia. The goal of unloading ships and marrying equipment with arriving units was achieved within ten days, and the first brigade (7th MEB) occupied its defensive positions within four days of its arrival.[1]

Existing MPF provides strategic and operational mobility and limited offloading capabilities absent a port. Typical MPF operations require ports and airfields to offload cargo, which makes the deploying force potentially vulnerable to enemy attack. The MPF concept was demonstrated in 1990 during Operation Desert Shield using a fixed port system. The Marine Corps armored vehicles aboard the MPF ships were the first heavy armor capabilities in that theater.

MPF (Future)

The MPF(F) squadron will be a single group of ships replacing one existing MPSRON.[2] The MPF(F) squadron (described in Appendix B) will be composed of five ship types loaded with the equipment needed to support a MEB. It is being designed to support an MPF(F) MEB of

---

[1]   Headquarters, United States Marine Corps, *Prepositioning Programs Handbook*, Washington, D.C., March 2005, p. 7.

[2]   The Marine Corps has stated a need for two MPF(F) MEB squadrons or one MPF(F) squadron plus two legacy MPSRONs. Ronald O'Rourke, *Navy–Marine Corps Amphibious and Maritime Prepositioning Ship Programs: Background and Oversight Issues for Congress*, Washington, D.C.: Congressional Research Service, RL32513, updated July 26, 2006, p. 18.

about 14,500 Marines. These ships will provide functions not provided by the MPF:

- At-sea arrival and assembly of expeditionary forces
- Interoperability with ESGs and CSGs
- Sea-based sustainment of expeditionary forces
- At-sea reconstitution and redeployment of the expeditionary force.[3]

An MPF(F) squadron will include equipment, such as rotary wing aircraft and surface connectors, vital to logistic support. So equipped, the MPF(F) squadron is referred to as a Maritime Pre-positioning Group (MPG).

Under Sea Basing logistics concepts, MPF(F) will deliver cargo to improved ports or over the beach in support of MAGTFs ashore. Maintenance, repair, medical treatment, and supply operations will be conducted primarily from sea-based platforms. The logistics infrastructure will be supported by the MPF(F) and will be maintained afloat and replenished from ships arriving on station from the continental United States (CONUS) or from support bases located nearer the operation.

Current plans call for an MPF(F) squadron to consist of two LHA Replacement (Amphibious Assault Ship, general purpose; LHA(R)) large-deck amphibious ships, one Amphibious Assault Ship, multipurpose (LHD) large-deck amphibious ship, three dry cargo/ammunition (T-AKE) ships, three Large Medium Speed Roll-on, Roll-off (LMSR) cargo ships, three Mobile Landing Platform (MLP) Landing Craft Air Cushion (LCAC) transport ships, and two legacy "dense pack" MPF ships taken from an existing squadron. These ships are described in Appendix B.

Sea State Considerations

Several technical challenges are inherent in the MPF(F) concept. Perhaps the most critical challenge is the difficulty of ship-to-ship transfer in high sea states, which will require precise positioning of ships. Pre-

---

[3] "Support Ships," *PEO Ships*, 2007.

cise positioning may also be needed to provide leeward protection for MLPs, as shown in Figure 2.1. Transfers of heavy loads using cranes in high sea states will additionally require new capabilities to compensate for relative motion between ships and the tendency of crane cargoes to swing.

As part of its MPF(F) research and development program, the Program Executive Office, Ships, assessed technology for automated ship heading and position control. Such systems were found to have low technical risk; they are now in commercial use. Further, a Low-Speed Roll Mitigation System that employs passive anti-roll tanks could increase large ship stability. It too is in commercial use.

Despite the above technologies, heavy load transfers between large ships and from large ships to MLPs remain a challenge. A ship bumper technology, Deep Draft Composite Fenders, for transfers between large ships, is now in development and has been tested at sea. It has a high technology readiness level. Commercial container ship carriers, such as Maersk, Ltd., and others, have successfully demonstrated stabilized crane technologies and open ocean fendering systems

Figure 2.1
Test for Motion Reduction in Lee of Cargo Ship

SOURCE: ⊠Support Ships,⊠ *PEO Ships*.

RAND *MG649-2.1*

that permit transfer of International Standards Organization (ISO) containers and even larger loads in heavy sea conditions. The problem of transferring heavy loads between large ships is therefore manageable and should be solvable without a large and/or difficult development program.[4] Stabilized crane technology is being improved, but is still limited in capability.[5]

A threshold of Sea Base operability through Sea State 3 (associated with wind speeds of 7 to 10 knots, or 8 to 12 miles per hour, with waves about 2 feet high) has been set. An objective of operability through Sea State 4 (associated with winds of 11 to 16 knots, or 13 to 18 miles per hour, with waves about 3 feet high) has been set. Table 2.1 shows the frequency of occurrence for Sea State 3 conditions over various regions.[6]

Table 2.1
Percentage of Sea State 3 or Less Conditions for Various Littoral Regions

Western Atlantic	60	Mediterranean Sea	75
Eastern Atlantic	40	Persian Gulf	89
North Sea/English Channel	52	North Arabian Sea	73
Eastern Pacific	45	West Indian Ocean	52
West and So. Caribbean	53	Cape of Good Hope	21
Northeast South America	54	Gulf of Guinea	71
Western South Atlantic	43	Northwest Africa	48
Eastern South Pacific	40	East Coast of Japan	48
Northwest South America	55	East Coast Philippines	62
Western Central America	73	Korean Coast	71

---

[4] Naval Research Advisory Committee, Panel on Sea Basing, *Sea Basing*, Washngton, D.C.: Office of the Secretary of the Navy (Research, Development and Acquisition), March 2005, p. 37.

[5] Defense Science Board, Task Force on Mobility, *Enabling Sea Basing Capabilities*, Washington, D.C.: Office of the Under Secretary of Defense for Acquisition, Technology, and Logistics, September 2005, p. 60.

[6] Defense Science Board, Task Force on Mobility (2005, p. 37).

Using the threshold value of Sea State 3, this table suggests that undegraded logistics operations from a sea base will be possible at least 70 percent of the time in the high profile regions of the Persian Gulf and North Arabian Sea, the Mediterranean Sea, the Gulf of Guinea, and the Korean Coast.

Sea Basing Operational Concepts

This section examines conceptual issues identified as part of this study. It first highlights key elements of Marine Corps concepts regarding use of the sea base—specifically, the MPF(F). It then examines key Army concepts. Finally, we introduce the three operational scenarios used later in the analysis.

Marine Corps Concepts

The Marines regard the MPF(F) as a major step forward in their ability to operate from the sea under the rubric of Operational Maneuver from the Sea. Today's Maritime Pre-positioning Ship Squadrons (MPSRONs) require safe, usable ports in order to offload cargo. Additionally, today's MPSRON ships are loaded in a "dense pack" configuration, which means that several days of work at or near the SPOD are required before the MEB equipment carried aboard the MPSRON is operational. While MEUs can deploy and sustain from their three-ship Expeditionary Strike Groups, the MEU is a battalion-sized task force. The MPF(F) will give the Marines the ability to deploy and sustain an entire brigade (less its fixed-wing fighters) from the 14 ships of the squadron.

Discussions with Marine Corps Combat Developments Command (MCCDC) revealed that the Marines' preference is to logistically support the MEB, once it is ashore, via cargo-carrying aircraft (MV-22 and CH-53K). This allows the MEB to (1) avoid creating a traditional "iron mountain" of shipborne supplies and material on the shore, and (2) facilitates the MEB's rapidly maneuvering inland once ashore. Additionally, the Marines want to retain several MV-22s on the sea base for casualty evacuation (we accordingly dedicated MV-22 aircraft and associated deck spots in our analysis).

The Marines also envision that some number of the available MV-22 sorties (and possibly some of the CH-53K sorties) would be used for tactical mobility missions for the forces ashore.[7] For example, depending on the tactical situation, the MEB commander might want to use some of the aircraft missions to conduct air assaults by company or battalion-sized forces. In terms of our analysis, the identification of "excess" air sorties (MV-22 and/or CH-53K) could be interpreted as the ability (or not) of the sea base to simultaneously provide logistic support to Marine Corps and Army forces ashore, while retaining for the MEB commander the capability to conduct other maneuver-related air missions.

Current plans envision the replacement of one of the three existing MPSRONs by an MPF(F) squadron. In a future crisis requiring multiple brigades, it is likely that a combination of ESGs and the MPF(F) squadron would form the initial Marine Corps force. The traditional "dense packed" MPSRON would arrive later, if needed, to bring the Marine Corps force ashore to division, or larger, size. Meanwhile, some combination of Army brigades might also arrive.

The Marines envision operating a considerable number of the MEB's aircraft from the sea base. However, the three large flight decks of the planned MPF(F) squadron are not sufficient to allow the Joint Strike Fighters (JSFs) of the MEB's air element to conduct sustained operations from the sea base (small numbers of JSFs could, however, use the MPF(F) as a base for refueling and for rearming or emergency landings). This is an important issue, in terms of the Army's concepts for at-sea basing of its own aircraft. The next section elaborates on this issue.

The Marines see the primary purpose of the MPF(F) as being to support the operations of the MEB. A recently concluded Analysis of Alternatives for the MPF(F) considered a MEB assault conducted from MPF(F) ships, followed by sustainment of the MEB from the same MPF(F) ships. Indeed, the MPF(F) as envisioned will be loaded with

---

[7] These preferences are reflected in our analysis as rules and data.

the initial supplies and equipment of a MEB. In terms of our analysis, we always assumed that the MPF(F) would support the MEB as its first priority. Once that mission was accomplished, any "excess capacity" was identified as potentially available to support other joint forces—specifically, Army brigades of various types. Of course, successful sustainment requires that the sustainment needs of both the MEB and the Army brigade in question be met.

The Army and Marine Corps ground elements of interest are characterized in Appendix C; their sustainment requirements are described in Appendix D.

Army Concepts

From 1996 until roughly 2002, much of the Army's future concept development focused on deploying and sustaining the Army via inter- and intracontinental aircraft. Subsequently, the Army began to move away from the idea that considerable Army forces (i.e., multiple brigades) could be moved and sustained by air. The high cost of the number of aircraft required under the Army's concepts has forced the Army to increasingly move in the direction of deploying and sustaining its forces by sea—despite the fact that the Army's Future Combat System (FCS) is still being designed with airlift factors (vehicle size and weight) in mind.

Today, the Army increasingly favors deploying and sustaining its forces from the sea. In a real sense, the Army focus on deploying its forces by sea has deep historical roots: the Army has deployed the vast majority of its forces by sea in every major conflict since the Spanish-American War, including in Operation Iraqi Freedom. This Army move has, of course, implications for the roles and missions of the Army–Marine Corps relationship. Nevertheless, the Army's renewed focus on operations from the sea has substantial potential benefit for the Department of the Navy: the Army could become an advocate for increased shipbuilding budgets, for example.

The Army emphasizes deploying its forces directly into the operational area via High-Speed Ships rather than pre-positioning its forces forward. In this regard, the current Navy–Marine Corps sea basing concepts (centered on the MPF(F)) are not directly compatible with the Army's desires. However, very little money has actually been earmarked for the hypothetical large High-Speed Ship (HSS) that the Army wants.

The Army places much less emphasis than the Navy–Marine Corps on at-sea transloading of forces in the manner for which the MPF(F) is currently being designed.[8] This analysis suggests, however, that the capability to perform at-sea transfer of Army forces could greatly benefit the joint force. The quantitative section of this study provides the detailed results, but as a preview, the analysis indicated that an Army Stryker Brigade (with about 15,000 tons of supplies and equipment) or heavy brigade (with about 20,000 tons of supplies and equipment) could arrive at the sea base and be moved ashore in 2 to 6 days, depending on such key variables as the distance offshore, the level of combat to be sustained, the availability of a Joint High-Speed Vessel (JHSV) to supplement the LCACs organic to the MPF(F), and prevailing sea states. That finding represents a new capability for Army forces.

As noted above, however, the current configuration of the sea base, with three large flight decks, limits the large-scale use of the sea base by Army aircraft. Until and unless most of the MEB's aircraft move ashore, or have another Navy ship as a base, there simply will not be room on the MPF(F) for significant numbers of Army aircraft. Our analysis does, however, show that there will generally be sufficient space aboard the three large flight decks of the MPF(F) to permit a small number of Army aircraft (roughly 1–2 deck operating spots per ship) to use the sea base on a temporary basis. Another important consideration regarding Army aircraft being based on the MPF(F) is the fact that most Army aircraft are not built for shipboard use—their blades do not fold automatically, and they lack braking systems. Additionally, few Army pilots are qualified to conduct landings on moving ships. In light of recent Army and Air Force helicopter operations during contingencies in Grenada, Panama, Somalia, and Haiti, these shortcomings are obviously not disqualifying.

---

[8] *Transloading* involves ship-to-ship movement by ramp. Transloading operations are illustrated in Figures B.3 and B.4.

Note that the Army has three distinctly different types of brigades: light (including airborne), Stryker (generally considered medium forces, since its armored vehicles are in the 20-ton class and are wheeled as opposed to tracked), and heavy (armed with M-1 Abrams–series main battle tanks, Bradley infantry fighting vehicles, and self-propelled artillery). Whereas Marine MEBs are generally similar, the weight (tonnage) and daily logistics requirements of the three different types of Army brigades vary widely.[9]

Operational Scenarios

We developed three operational scenarios for this analysis. These scenarios are intended to represent the most likely cases for which conventional Army forces (airborne, Stryker, and heavy brigades) could interact with a sea base. All cases are in the context of a Major Combat Operation in which the major elements of a MEB have gone ashore, are in combat, and are being sustained by the sea base as Army forces are introduced. With the MEB established ashore, the threat to the sea base might plausibly be reduced. In no case were Army aircraft included as lift assets; the Army brigade was considered to have all its normal organic assets other than aircraft.[10]

Scenario A—Army Forces Arrive Inland

In this scenario, it is assumed that an Army light or airborne brigade arrives 50 to 75 nautical miles (NM) inland, possibly as part of a joint forcible entry operation, soon after the MEB's arrival ashore.[11] Two

---

[9] Appendix D provides logistics data for Army and Marine Corps ground elements.

[10] Depending on the situation, the Army envisions that considerable numbers of Army aircraft (UH-60 or CH-47 cargo helicopters, and AH-64 Apache attack helicopters) might be temporarily located on the sea base. The Army feels that sea basing its aircraft could greatly increase the combat power of the initial Army forces deployed ashore. In consequence, this analysis considers the feasibility of placing a significant number of Army helicopters on a sea base for some time.

[11] With Army forces 50 to 75 NM inland, sustainment from the sea will be from greater distances. We consider aerial sustainment distances of 75 to 110 NM.

main cases are considered in this scenario. The first main case is consistent with the Marine Corps' preference for aerial sustainment. Here, both the Army light or airborne brigade and the MEB are sustained entirely using MV-22 and CH-53K aircraft from the sea base. The LCACs of the MPF(F) are not utilized in this case (perhaps because both the Marine Corps and Army forces are so far inland that they can no longer benefit from supplies deposited at the beach by the LCACs).

In the second main case, the MEB can use LCACs to sustain it through a beach or SPOD.

Scenario A is particularly stressing—so much so that this analysis considers means for enhancing sustainment from the sea base. Key features of Scenario A are depicted in Figure 2.2.

Scenario B—Army Forces Enter the Area of Operations Directly

This scenario represents the Army's preferred option. Today, using LMSRs, or in the future possibly using HSS vessels, Army forces would

Figure 2.2
Operational Scenario A

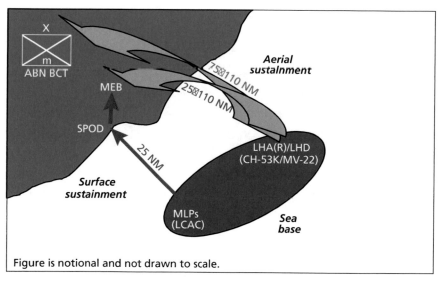

Figure is notional and not drawn to scale.

move directly to a usable port, offload, and then start operations ashore as soon as possible. In this scenario, we examined the ability of the sea base to simultaneously support both the MEB and either a Stryker or a heavy brigade from the Army.

The logistics requirements of these Army brigades are much greater than those of a light brigade because of the higher fuel requirements of armored vehicles and the heavier ammunition that these brigades use compared with a light force (e.g., 155mm howitzers firing 100-pound shells compared with 105mm weapons firing 33-pound shells).[12]

A key variable examined in this scenario was the utility of LCACs as part of the resupply effort. As observed earlier, the Marines prefer that, once ashore, the MEB is resupplied to the maximum extent possible by aircraft flying from the sea base. In Scenario B, we examine that case as well as the case of adding LCACs to the logistics flow. In the latter case, it was assumed that the MEB and the Army brigade are (1) either close enough to the coast that it would be easy to pick up supplies delivered to the beach by LCACs or (2) the units were fairly deep inland (25 miles or more) but had the ability to send trucks to the beach to pick up those supplies delivered by LCACs.[13]

This assumption that the MEB, as well as Army forces being supplied by the sea base, could pick up LCAC-delivered supplies for movement inland by ground transport presumes that the routes from the beach (or small port that U.S. forces have access to) to the units operating inland are relatively safe. This may not always be the case, thus requiring the ground units to escort their supply vehicles and provide protection for the offload points at the beach or port.

Note that we did not envision a large amount of infrastructure being built to support operations at the beach—certainly nothing like the "iron mountains" associated with World War II–type amphibious operations. Sustainment operations would instead maintain only several days of supplies ashore. Nevertheless, the MEB commander, the affected Army commanders, and the Joint Force commander would

---

[12] See Appendix D, Sustainment Requirements, for additional information.

[13] See Appendix C, Army and Marine Corps Ground Elements Evaluated, for additional information.

have to accept the implications of cross-beach supply. The downside could be the need to provide protection and escort for the supplies arriving at and moving forward from the beach. The advantage is that, if LCACs are used to supplement the aerial delivery of supplies from the MPF(F), the amount of tonnage that could be moved is increased significantly.

Note also that, even if aerial resupply alone is being used and the area between the shoreline and the units operating inland is not completely secured, the resupply aircraft would also be at risk to enemy fire as they pass over the unsecured area en route to deposit their supplies at inland locations.

Finally, note that we did not analyze the number of trucks that would be required for the forward movement of supplies delivered to the beach by LCACs. It was assumed that sufficient numbers of supply trucks (including trailers) would be available to the Army and Marine Corps units operating ashore. A detailed examination of this issue, which was beyond the purview of this study, should be conducted as part of follow-on analyses.

It could be argued that, if a port were available for the arrival of Army forces via LMSR or HSS, the sea base would not be needed to provide logistic support for Army forces. One plausible scenario is that the port facility is disabled by an enemy attack (e.g., a chemical weapons strike) after the Army force arrives at it. Another plausible scenario is that, following its arrival, the Army brigade rapidly advances along the coast away from the port by which it entered, eventually getting much closer to the location of the MEB/sea base, at which point the sea base would assume responsibility to support the Army brigade as well as the MEB.

The situation in Scenario B is depicted in Figure 2.3. Although the diagram below shows the Army brigade being farther inland than the MEB, that would not necessarily be the case in an actual operation. The MEB could be deeper inland than the Army unit at the time the Army forces come under the purview of the sea base for logistic support.

Figure 2.3
Operational Scenario B

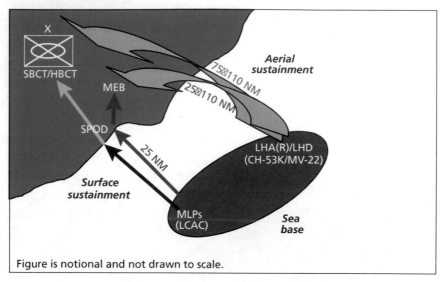

Figure is notional and not drawn to scale.

Scenario C—Army Forces Enter the Area of Operations via the Sea Base

The Army uses the sea base in Scenario C to transload, at sea, an Army brigade that is then moved ashore by LCACs (or, in some excursions, LCACs and a JHSV) and, to a lesser extent, by CH-53 and MV-22 aircraft—a natural ship-to-shore movement for the Navy–Marine Corps team since World War II, but much less common for the Army. As mentioned in Scenario B, the Army's preference is to deploy directly into a usable port via High-Speed Ships. Army forces rarely practice transloading troops and equipment at sea. This scenario is important because it shows how the MPF(F), as conceived by the Department of the Navy, could introduce an important new capability for the Army. In this scenario, no usable port may as yet be available to the joint force commander, who wants to rapidly introduce Army medium or heavy forces ashore to supplement the MEB that is already fighting there. Rather than waiting for the seizure (and possible repair) of a port capable of accepting LMSRs or HSSs, this option would give the joint force commander the ability to introduce an Army brigade ashore via the sea base.

As in Scenarios A and B, the MEB is assumed to be ashore, with the sea base providing its logistic support. While the MEB is engaged in operations, an Army Stryker or heavy brigade arrives at the sea base. Importantly, it is assumed that the Army ships are loaded in a way that allows selective offload of equipment via ramps onto the three Mobile Landing Platform ships of the squadron. If the Army ships are dense packed, they might not be able to capitalize on this capability.[14] Additionally, it is assumed that Army personnel will have received sufficient training in at-sea transfer operations to make the mission feasible.

The Army brigade's equipment and personnel are transloaded from Army shipping onto an MLP and then ashore via the LCACs of the squadron. It would be advantageous here for most Army personnel to travel ashore in the LCACs at the same time as their vehicles, thus facilitating maintenance of unit integrity as the brigade builds up ashore. In some excursions, a JHSV was added to supplement the LCACs. The concept here is that at least two JHSVs would be used to bring troops into theater. Once in theater, one JHSV would be used to help move troops to the sea base (possibly from an intermediate staging base) while a second JHSV moves Army personnel, supplies, and equipment ashore from the sea base.

The scenario is diagrammed in Figure 2.4. Note that although the diagram includes an SPOD, the actual debarkation of Army forces would likely be accomplished by LCACs landing at a beach. When a JHSV is included, a small port would, of course, be required. In that case, the LCACs may be able to deposit their loads over the beach, while the JHSV enters what may be a fishing village–sized port to offload its cargo and passengers.

---

[14] To access items of interest, selective offload will be accomplished by moving cargo internally. The storage efficiency of ships capable of selective offload will be less than that of dense packed ships, which are loaded to maximize storage efficiency. The Army would need additional pre-positioning ships to achieve capability for selective offload.

Figure 2.4
Operational Scenario C

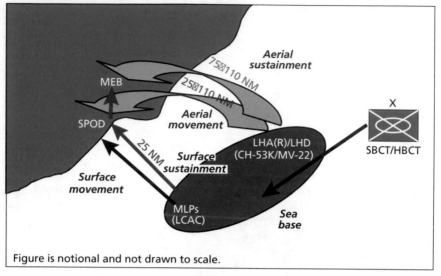

Figure is notional and not drawn to scale.

CHAPTER THREE

Scenario Analysis

In examining Department of the Navy Sea Basing analyses, we initially found a seeming disconnect between analyses conducted by the Strategic Mobility and Combat Logistics branch of OPNAV (N42) and by MCCDC. N42 analyses, conducted with modeling support from the Center for Naval Analyses (CNA) and SRA International, concluded that intertheater, intratheater, intra–sea base, and tactical re-supply capabilities under sea basing concepts were adequate to sustain multiple brigades.[1] The MCCDC analysis was prepared for the Capabilities Development Document (CDD) analysis in preparation for an MPF(F) Analysis of Alternatives (AoA).[2] In the scenario that MCCDC examined, one Sea Base Maneuver Element (SBME)[3] is sup-

---

[1] An N42 National Defense Industrial Association (NDIA) 2004 Joint Seabasing Logistics briefing presented in October 2004 (Jonathan Kaskin, "Seabasing Logistics CONOPs," briefing to NDIA 10th Annual Expeditionary Warfare Conference, October 2004) concluded (slide 19) that less than 40 percent of the MPF(F) ships' assets and helicopter spots would be used for Marine Corps MEB sustainment. The analysis points to potential excess capacity to support joint sustainment, and illustrates potential capability with a Maritime Pre-positioning Group (i.e., an MPF(F) squadron, together with air and surface connectors needed to conduct logistics operations) supporting a MEB, a Stryker Brigade Combat Team (SBCT), and Special Operations Forces (SOF) simultaneously.

[2] MCCDC, Mission Area Analysis Branch, "MPF(F) CDD Analysis: Results for Seabasing Capabilities," briefing, March 23, 2006a.

[3] The MEB designed for MPF(F) operations, referred to as the MPF(F) MEB, is composed of a Shore Base Echelon (SBE), a Forward Base Echelon (FBE), and a Sustained Operations Ashore Echelon. Within the SBE are the Sea Based Maneuver Element (SBME), that portion of the SBE that is projected ashore for operations and its support element, and the Sea Base Support Element (SBSE). The FBE is made up primarily of fixed wing assets organic

ported with some difficulty from MPF(F) ships.[4] Recognizing that differences in scenarios and assumptions existed between the two studies, we used the MCCDC analysis as a starting point for a broader examination of factors related to successfully sustaining more than one brigade ashore.

For the Army brigades (light/airborne, Stryker, or heavy), we examined "pure" brigades—not including other units that would normally accompany a brigade into action. For example, no aviation or extra supply units were included in the brigade. We recognize that the Army would want to introduce these elements as quickly as possible after the arrival of the brigade combat team. In many respects, the supply throughput capacity of the sea base is providing most of the logistics needs of the brigades, thus reducing the need for divisional-level support units to accompany the Army unit, at least for the first few days of operations. Additionally, we assumed that the logistics needs of the Army units would be met by supply ships that would arrive at the sea base, loaded with Army supplies, thus minimizing the need for the Army units to have to rely on the MEB's supplies, which are already loaded aboard the MPF(F) ships.

The initial step in our quantitative analysis was to redo the CDD analysis using a simulation (described in Appendix E) developed for this study.

The MPF(F) CDD analysis examined sustainment from MPF(F) ships using only rotary wing (CH-53K and MV-22) aircraft.[5] Sustain-

---

to the MEB, such as the KC-130 and EA-6 squadrons and their support; its elements will self-deploy to a forward operating base. The Sustained Operations Ashore Echelon normally remains in CONUS. The SBME and the entire SBE (i.e., the SBME and the SBSE) are the only portions of the MEB that might be sustained ashore from the sea base. This study considers sustainment operations for the SBE in heavy combat and in sustained combat operations, as well as for the SBME in heavy combat operations.

[4] Difficulty in sustaining the SBME using only CH-53K and MV-22 aircraft is illustrated by the CDD analysis, which found that an SBME cannot be sustained within a period of darkness using procedures optimized to do so.

[5] The CDD analysis considered both assault and sustainment from MPF(F) ships. It included ship-to-shore movement over 25 NM, with the landing team moved to the sea base before the assault and launched from it. Movement was accomplished using 48 MV-22 and 20 CH-53K aircraft, and 18 LCAC surface connectors. Taking into account operational

ment was to be provided from a distance of 110 NM and during a single period of darkness (eight to ten hours). MCCDC supported our study by providing sustainment rates and lift capacities for CH-53K and MV-22 aircraft having internal and external loads. CNA provided additional data. With these data, but using a RAND-developed simulation, we arrived at a conclusion similar to that reached in the CDD analysis: that an SBME can be sustained with some difficulty at a distance of 110 NM from a sea base.

In this analysis, we categorized sustainment requirements in the same way that MCCDC did for the CDD analysis, and we used the same number of lift assets.[6] For presentation purposes, our analysis consolidates all sustainment and lift requirements using the simple metric of tons per day. The model developed for this study operates sustainment assets at full capacity for indefinite sustainment (i.e., at a pace that can be maintained for a considerable period as opposed to surge operations, which can be maintained for only a few days).

We analyzed distances of 25 to 110 NM from the large-deck LHA(R)/LHD ships to the SBME. The results, which are shown in

---

availability and the need to withhold MV-22 aircraft for missions such as search and rescue, the CDD analysis employed 34 MV-22 and 16 CH-53K aircraft, and 17 LCACs. These same aircraft, but not the LCACs, were used in sustainment.

[6] Sustainment requirements are categorized as follows: ammunition, dry stores, bulk Petroleum Oil and Lubricants (POL), and bulk water. Both analyses used the elements of the 2015 MEB Air Combat Element: 48 MV-22 and 20 CH-53K aircraft, plus 6 unmanned aerial vehicles (UAVs). The operational availability of MV-22 aircraft was taken to be 82 percent; we withheld five operationally available MV-22 aircraft for casualty evacuation and other missions (for a total of 34 MV-22 aircraft used in sustainment). Operational availability of the CH-53K was taken to be 80 percent. We withheld no CH-53K aircraft for other missions, so that a total of 16 CH-53K aircraft are used in sustainment. The operational availability of LCACs that have undergone a service life extension program was taken to be 95 percent, a significant improvement over the current LCAC.

With 95 percent availability and 18 LCACs on the MLPs, 17 LCACS are therefore used in sustainment. This matter requires some additional discussion. The historical rate at which LCACs lose operational availability has been about 6 percent per day. For example, if 17 LCACs are operationally available on a given day, it would be expected that only 16 LCACs would be operationally available the next day, and so on. However, future LCACs are expected to be more reliable than existing LCACs. Moreover, the MLP and its LCACs cannot be viewed as a closed system; the MPF(F) LHD can carry three LCACs and has a substantial capability to maintain and repair LCACs.

Figure 3.1, suggest some difficulty in sustaining an SBME conducting heavy combat operations using only CH-53K and MV-22 aircraft from a distance of 110 NM.[7] Fewer sorties, with smaller payloads, occur as distance increases. Our analysis further suggests that these aircraft alone cannot sustain an entire Sea Base Echelon (SBE) from a

Figure 3.1
Required Tons per Day and Lift Capacities, VTOL-Only Sustainment of the MEB

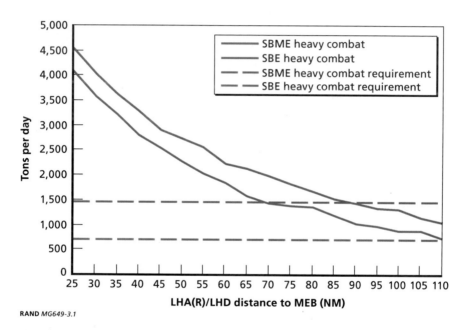

RAND *MG649-3.1*

---

[7] Presentations using tons per day of lift capacity as a metric can oversimplify results in some regards. The task of moving a ton of bulk liquid is different from the task of moving a ton of ammunition. There is also the factor of distance. For example, moving a ton of ammunition 25 NM is not the same as moving it 75 NM—at longer distances, payloads are reduced as fuel requirements increase and, with longer flight times, fewer sorties can be generated. These graphs reflect the differing sustainment requirements shown in Appendix D. Maximum lift capacity per day differs with differing constraints on those sustainment operations.

distance of 110 NM; the maximum range for which such sustainment is possible appears to be about 70 NM.[8]

The above results can also be presented using the metric of *relative lift capacity*, defined as the ratio of maximum sustainment capacity (in tons per day) to average sustainment requirement (also in tons per day). This metric can be viewed in several ways:

- Relative lift capacity reflects the robustness of available lift resources. As background, both the MPF(F) Analysis of Alternatives and this study assume favorable operating conditions, but they recognize that high sea states and other factors can degrade sustainment performance. High sea states hinder ship-to-ship transfer, and they slow and reduce the capacity of LCACs.[9] Other possible factors include the loss of aircraft. In light of the possibility of degraded sustainment capacity, a sustainment force that can provide little more than a required level of sustainment under favorable conditions offers no hedge against operational degradation. Given a periodically degraded sustainment capability, high relative lift capacity, exploited under favorable conditions, can offset operational degradation experienced under unfavorable conditions. Under this concept, sustainment assets attempt to maintain a fixed number of Days of Supply (DOS) for the ground elements.

- Relative capacity also reflects the flexibility of the sustainment force under favorable conditions. A sustainment force that can provide more than the required level of sustainment can spare assets (such as MV-22 aircraft) for use by ground elements. Similarly, such a sustainment force can meet sustainment requirements despite aircraft losses.

---

[8] The use of an entire SBE ashore is a worst case for this analysis. It serves to illuminate the limits of sustainment and failure modes in sustainment.

[9] Ship-to-ship transfer capability at the sea base is stated in terms of significant wave height. When all wave heights are measured (peak to trough), the *significant wave height* is defined as the mean value of the highest one-third waves. Ship-to-ship transfer is considered undegraded for significant wave heights of no more than three feet, or NATO Sea State 3.

- In analytic terms, high relative lift capacity is a hedge against analytic uncertainty; this analysis deals with notional platforms (such as MLPs and the JHSV) or platforms still in design (such as the LHA(R)[10] and the CH-53K helicopter), whose performance is uncertain. LCACs will undergo service life extension programs before MPF(F) ships enter service and will be replaced in the period of interest, making future LCAC operating characteristics uncertain.[11] Recognizing these and other uncertainties, we conclude that high relative lift capacity provides a margin for error in performance estimates.
- Finally, this metric can help identify and compare factors useful in achieving robust sustainment capability. For example, lift capacity metrics, such as tons per day, do not readily provide insight into the benefits of reducing lift demand. The relative capacity metric provides for direct comparisons in this case.

Again, the relative capacity metric is the maximum throughput capacity (in tons per day) divided by the sustainment requirement (also in tons per day). The results shown in Figure 3.1 are shown again, using relative capacity, in Figure 3.2 to illustrate that metric. Because sustainment is by air only, the distance from the MLPs to the MEB is irrelevant here; distances are from the large deck MPF(F) ships.

It appears just possible to sustain a single ground element when maximum sustainment capacity is equal to the required sustainment level—i.e., their ratio is 100 percent. Results shown in Figure 3.2 suggest that maximum lift capacity is about 130 percent of the

---

[10] The LHA(R) might prove to have a smaller aircraft capacity than it is credited as having. If so, the number of aircraft for sustainment would have to be reduced.

[11] An MCCDC, Mission Area Analysis Branch, analysis of surface assault connectors, completed in April 2006 ("Surface Assault Connector Requirements Analysis Update: Overview to Inform Seabasing Capabilities Study," briefing, April 13, 2006b) considered numerous possible sets of characteristics for an LCAC replacement. LCACs that have undergone service life extension are assumed here to have a maximum load capacity of 72 tons and a deck space of 1,809 square feet, and to average 35 knots in operation—consistent with the MCCDC analyses. The NRAC (2005) study of sea basing notes that LCAC speed and range are strongly affected by sea state.

**Figure 3.2
Relative Lift Capacities in MEB, VTOL-Only Sustainment**

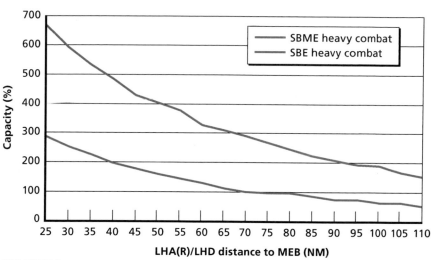

RAND *MG649-3.2*

SBME sustainment requirement from a distance of 110 NM. SBME sustainment then appears possible, but with little margin for operational degradation: few air assets are available for use by ground forces, and there is little leeway for uncertainty.

Our initial analysis suggested operational factors useful for enhancing sustainment capacity or for projecting an Army ground element more quickly from the sea base. We selected the following four options for enhancing sustainment capacity for analysis:

- Reducing distances from the large-deck MPF(F) and MLP ships to supported ground elements or SPODs. The significance of this factor was illustrated above. Of course, threat conditions can limit these distances; other options are needed.
- Adding LCACs to CH-53 and MV-22 aircraft in sustainment. LCAC connectors from MLPs are an attractive addition to rotary wing aircraft here. These LCACs were used in the MEB assault, but they represent an unused resource after the assault.
- Increasing the ratio of CH-53K to MV-22 aircraft. The aircraft mix used for sustainment in the MPF(F) AoA reflects the

need for mobility rather than for sustainment. In particular, CH-53K aircraft carry more than twice as much cargo as the MV-22 and are equally fast on ingress (external loads limit both aircraft to the same flight speed).[12] Increasing the ratio of CH-53K to MV-22 aircraft, seen to enhance sustainment, would be expected to enhance sustainment throughput.

- Reducing sustainment requirements. Reducing demand for external sustainment might enable sustainment of larger forces or sustainment of a given force at greater distances. For example, the U.S. Army is creating units to make brigades self-sufficient in bulk water. For perspective, on average, an Army airborne brigade in heavy combat consumes about 150 tons of water per day; an SBME in heavy combat consumes about 130 tons of water per day.

Our mobility analysis of the Army ground element in this study is patterned after the assault analysis in the MPF(F) AoA for consistency. In this analysis, we considered two new factors for improving performance:

- Increasing the ratio of CH-53K to MV-22 aircraft. Little benefit was expected from increasing the ratio of CH-53K to MV-22 aircraft. The real issue here is assurance that such a change would not degrade Army ground element movement time in Scenario C.
- Adding JHSVs to LCACs as surface connectors. A single JHSV about equals the combined lift capacities of LCACs from the sea base. The JHSV is also faster than the LCACs, which suggests that adding a JHSV to LCACs is an attractive option.[13]

---

[12] The MV-22's main advantage here over the CH-53K is its higher egress speed. In terms of moving a ground element, the MV-22 also has speed and survivability advantages in ingress.

[13] The Naval Research Advisory Committee (2005, p. 3) states the value of JHSV (generically, high-speed surface connectors) strongly:

> A high-speed surface connector (HSC)—a vessel that can move troops and materiel between the Sea Base and waters immediately offshore—will prove to be a critical enabler of Sea Basing. The HSC is essential to our ability to establish the Sea Base at a

These operational factors are explored in the following analysis of our three scenarios.

Scenario A—Army Forces Arrive Inland

Reducing Distances

Previous results indicated that, by reducing the distance from the sea base to an SBE in heavy combat, the SBE could be sustained using only CH-53K and MV-22 aircraft. Our analysis suggests that both an SBE (or an SBME) and an airborne brigade in heavy combat could be sustained simultaneously at shorter distances using only CH-53K and MV-22 aircraft.

The capability to sustain both ground elements simultaneously (Figure 3.3) at shorter distances[14] is considered marginal in the context of uncertainties and the potential for performance degradation through factors such as high sea states or aircraft losses.

Adding LCACs

We now turn to a second means of increasing sustainment capacity: using LCACs as additional connectors. Here, we see a more robust sustainment capability—enough to provide a significant hedge against operational uncertainties and potential performance degradation.

---

secure stand-off distance. We see no realistic near- or mid-term alternatives to an HSC if the Sea Base is to have the capability of moving heavy materiel—in particular armored combat vehicles—to forces ashore. A properly designed HSC will afford important synergies with the legacy landing craft air cushion (LCAC), which we also regard, for all its limitations, as an indispensable system offering unique heavy-lift capabilities over the beach.

[14] In Scenarios A and B, the distance from large deck MPF(F) ships to the Army ground element is assumed to be 50 NM greater than the distance from the sea base to the Marine Corps ground element. Distances from the sea base to ground elements are then paired as shown in Figure 3.3.

Figure 3.3
Scenario A, VTOL-Only Sustainment of a MEB and an Army Airborne Brigade, Is Marginal

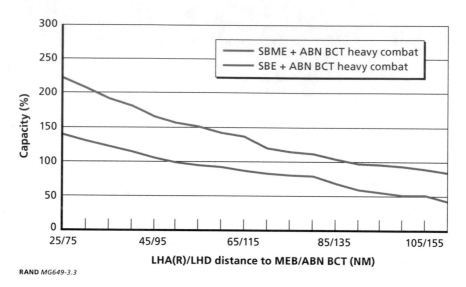

RAND *MG649-3.3*

Sustainment performance using LCACs along with rotary wing aircraft is shown in Figure 3.4 (solid lines) and compared with the above result (dashed lines).[15] However, some of the robustness shown here is illusory: the additional capacity provided here by LCACs directly benefits only the MEB; the BCT benefits only indirectly as rotary wing aircraft, no longer needed for MEB sustainment, become available for BCT sustainment. A breakpoint is reached when the air assets cannot sustain the BCT (regardless of total sustainment capacity).[16]

The situation is illustrated in Figure 3.5, which shows the sustainment levels for the SBE and the BCT separately.[17] This figure shows

---

[15] In Scenarios A and B, the distance from the MLPs to the SPOD used for sustainment is taken to be 25 NM.

[16] Sustainment breakpoints occur only in Scenario A; LCACs augment MPF(F) aircraft in Army sustainment in Scenario B.

[17] Irregularities in the curve for the Army data result from the assumption that the MEB has first priority in sustainment and preferences built into the model.

Figure 3.4
Scenario A, VTOL Plus LCAC Sustainment, Is More Robust

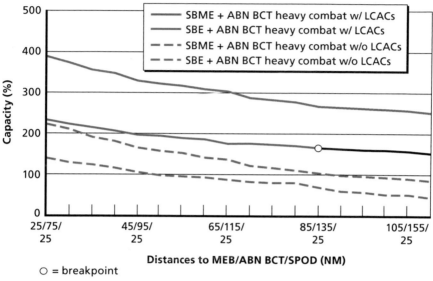

RAND *MG649-3.4*

a nearly constant level of sustainment for the MEB, reflecting LCAC sustainment from a fixed distance (25 NM).[18] It also shows sustainment to the BCT declining as LHA(R)/LHD distance to the BCT increases (again, sortie rates decline and aircraft payloads decrease) with the limit of BCT sustainability reached at a distance of about 85 NM. The circle in Figure 3.4 indicates this breakpoint.

As noted earlier, when LCACs were used to augment the movement of supplies ashore, it was assumed that the MEB would be able to pick the supplies up at the beach and move the supplies inland to the Marine units needing them.

---

[18] We examined the implications of using a fixed 25-NM LCAC movement distance and found them to be insensitive to this distance. Doubling the movement distance decreases throughput by about 15 percent, because increasing this distance (1) does not change CH-53K and MV-22 performance and (2) does not change LCAC load and offload times; however, the LCAC sortie rate is then reduced by 25 percent. See Appendix A for a fuller discussion of this matter.

Figure 3.5
Scenario A, Breakpoint in Army Sustainment

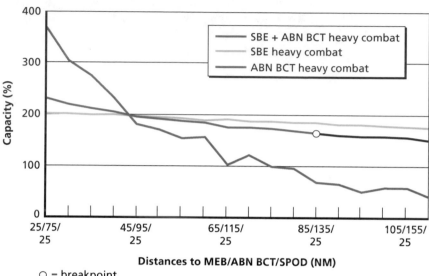

O = breakpoint

RAND *MG649-3.5*

Increasing the Ratio of CH-53K to MV-22 Aircraft

Changing the aircraft mix used to sustain Army and Marine Corps ground forces is a third potential means of increasing capability. Our analysis suggests that sustainment performance can be improved significantly by increasing the ratio of CH-53K helicopters to MV-22 aircraft.

The MCCDC analysis used a mix of 16 operational CH-53K and 34 operational MV-22 aircraft for sustainment. For this portion of the analysis, we reversed that ratio, to 34 operational CH-53K aircraft and 16 operational MV-22 aircraft for sustainment,[19] to illuminate how changing the mix of rotary wing aircraft aboard the MPF(F) ships can change sustainment performance. We are not proposing this as "the right mix" of aircraft.

---

[19] This value does not include the five MV-22 reserved for casualty evacuation and other missions. We did not consider aircraft size (spot factor) in this simplistic analysis.

This aircraft mix provides a more robust capability to sustain an SBME in heavy combat and an improved ability to sustain an entire SBE (Figure 3.6). However, it does not enable SBE sustainment from 110 NM using only rotary wing aircraft.

The benefits of reversing the mix of rotary wing aircraft for sustainment performance are comparable to adding LCACs as connectors. In combination with the addition of LCACs, this mix of CH-53K and MV-22 aircraft further increases sustainment capacity (Figure 3.7). New results are shown here with solid curves, and the results from Figure 3.4 are included as dashed curves, for comparison. As in that figure, airborne sustainment for Army ground forces can be limiting, but the breakpoint in Army sustainment can be pushed to greater distances by changing the aircraft mix.

Figure 3.6
Altered Aircraft Mix in Scenario A Gives More Robust Sustainment

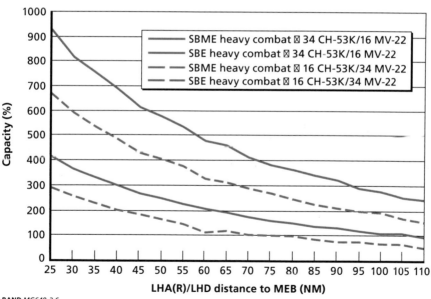

RAND *MG649-3.6*

Figure 3.7
LCACs Plus Altered Aircraft Mix in Scenario A Give Greater Robustness

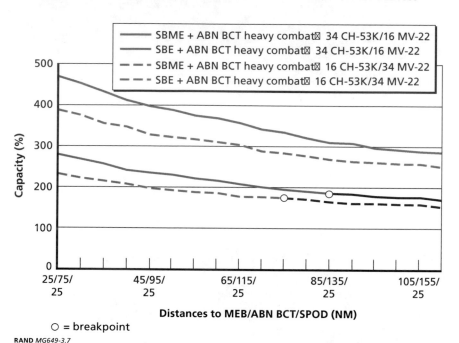

O = breakpoint

RAND *MG649-3.7*

Reducing Sustainment Demand

Eliminating bulk water requirements from the sea base illustrates the potential for reducing sustainment demand: it would significantly increase the capability for combined SBME and airborne brigade sustainment using VTOL and LCACs.

In some operational circumstances, eliminating or significantly reducing the requirement for water might be possible, if water sources are available ashore and efforts to purify water are included; in other situations, sources of potable water may not exist. Model results supporting this finding are shown in Figure 3.8, and the results shown in Figure 3.4 are included as dashed curves.

Freed of the requirement of sustaining brigade combat teams with water by air, better sustainment in fuel, ammunition, and dry stores

Figure 3.8
Scenario A Sustainment, Using LCACs with and without Bulk Water

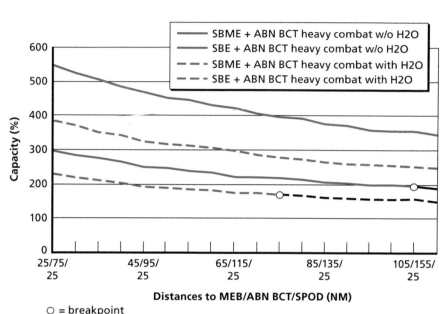

○ = breakpoint

RAND *MG649-3.8*

(food, consumables, and spare parts) can be provided to the airborne force.

Scenario B—Army Forces Enter the Area of Operations Directly

Scenario B differs from Scenario A in two primary regards. First, Scenario B entails sustaining Stryker or heavy brigades, which have higher sustainment requirements than the airborne brigade sustained in Scenario A. In tons per day, the SBCT has a sustainment requirement about 30 percent greater than that of the airborne brigade. The second main difference is that, whereas the burden of sustaining the Army brigade combat team fell entirely on CH-53K and MV-22 aircraft in Scenario A, LCACs directly assist in Army sustainment in Scenario B. LCACs increase the capability to sustain the Army ground element (a limiting factor in some cases), and they increase operational flexibility by improving the matching of connectors and payloads.

Our main finding is that the effects of increased ground element sustainment requirements are largely canceled by the greater operational flexibility in sustainment. The SBCT represents an increase of less than 10 percent over the airborne brigade in combination with the SBME—a marginally higher sustainment burden on the sea base. Similarly, the HBCT represents an increase of less than 30 percent over the airborne brigade in terms of overall daily sustainment requirements.[20]

As in Scenario A, the analysis began with consideration of sustainment performance with and without LCACs. Without LCACs, a performance reduction commensurate with 10 to 30 percent higher sustainment demand is seen. With LCACs, the ability to sustain both an SBME and an SBCT in heavy combat (shown in Figure 3.9) is similar to that seen for the SBME and an airborne brigade (shown previously in Figure 3.4). The effect of replacing an SBME with an SBE far exceeds that of replacing an airborne brigade with an SBCT.

As expected, performance worsens when the sea base must sustain either an SBE or an SBME, along with a heavy brigade, in heavy combat. As shown in Figure 3.10, sustainment of an SBE or an SBME with a heavy brigade is feasible with LCACs. Without LCACs, the ability to sustain both an SBME and an HBCT appears marginal at best. In addition, without LCACs, the sea base cannot sustain both an SBE and an HBCT.

Specific findings of our analysis of Scenario B are as follows:

---

[20] Appendix D describes and compares requirements for Army and Marine Corps brigade sustainment. Here, briefly, are the requirements: an SBME in heavy combat consumes on average 680 tons of bulk liquids, ammunition, and other supplies per day. An airborne brigade, also in heavy combat, consumes on average 299 tons per day for a total of 979 tons per day. An SBCT consumes on average 394 tons per day (or an additional 95 tons per day over that of the airborne brigade, increasing the total consumption rate by less than 10 percent). An HBCT consumes on average 583 tons per day (increasing the total for an airborne brigade by an additional 284 tons per day, increasing the total consumption rate by just less than 30 percent). The net affect of substituting an SBCT for an airborne brigade is thus less than 10 percent and the net affect of substituting an HBCT for an airborne brigade is less than 30 percent.

- Net sustainment requirements are increased less than 30 percent with an SBCT or an HBCT in place of an airborne brigade as an Army ground element.
- Increased flexibility in matching connectors with payloads largely offsets the additional sustainment demand seen above.
- The use of LCACs to sustain the Army ground element obviates the problem seen in Scenario A of the limitations of air-only sustainment of the BCT.

Scenario A is then seen as more stressing than Scenario B, so Scenario B is not analyzed as thoroughly as Scenario A.

Figure 3.9
SBCT Sustainment in Scenario B, with and without LCACs

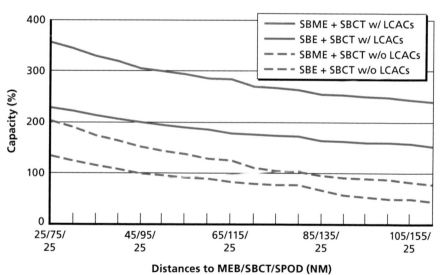

Figure 3.10
HBCT Sustainment in Scenario B, with and without LCACs

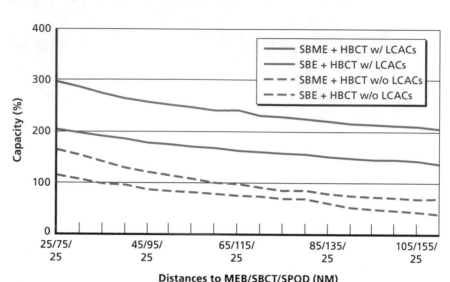

SBME + HBCT w/ LCACs
SBE + HBCT w/ LCACs
SBME + HBCT w/o LCACs
SBE + HBCT w/o LCACs

RAND MG649-3.10

Scenario C—Army Forces Enter the Area of Operations via the Sea Base

The MPF(F) Analysis of Alternatives considered a Marine Corps assault from MPF(F) ships, with those ships inserting an SBME under cover of darkness. The operational concepts employed in that analysis are used here. In particular, Army SBCT or HBCT personnel will be positioned on the MLPs for movement ashore, and movement will use LCACs and CH-53K and MV-22 aircraft.[21] This scenario assumes that Army LMSR ships are in theater and can immediately flow vehicles, ammunition, and dry stores onto the MLPs at least as quickly as connectors can take them ashore.

Aside from the insertion of Army ground elements in place of a Marine Corps ground element, there are three main differences between

---

[21] LCAC, CH-53K, and MV-22 operations are described in detail in Appendix D. In the base case, LCACs operate 16 hours a day with overlapping periods of operation for the MLPs. Similarly, large-deck MPF(F) ships have overlapping flight windows 10 hours long.

this scenario analysis and the AoA. First, in the MPF(F) AoA, there were no sustainment requirements during the Marine Corps assault; the diversion of significant lift assets for MEB sustainment is a clear impediment to force movement ashore. Second, the Marine Corps assault was conducted from a distance of 25 NM from the shore. With the expectation that Army BCT movement cannot be accomplished in a single cycle of darkness, our analysis considers force movement from distances of 25 to 50 NM from the objective area. Third, we consider as an excursion the use of a Joint High-Speed Vessel as an additional surface connector.

The performance metric for this scenario is the time to complete Army brigade movement. Here, the MEB is assumed to operate inland, and its sustainment is delivered 25 NM farther than to the Army objective area, from distances of 50 to 75 NM instead of 25 to 50 NM for the Army.[22]

As noted above, MPF(F) ships' aircraft, LCACs (and possibly a JHSV) are used in Army brigade movement. Army analysts have examined the transportability of SBCT and HBCT supplies and equipment by MV-22 aircraft and have found that MV-22 aircraft can transport the large majority of those supplies and equipment. Our examination of HBCT data indicates 1,770 vehicles plus an additional 1,957 tons of miscellaneous equipment need to be moved. Many of the lighter vehicles are trailers and light trucks (weighing about 2 tons each). CH-53K aircraft could transport 1,396 of the HBCT's 1,770 vehicles a distance of 110 NM. JHSVs and CH-53K and MV-22 aircraft would be used to move lighter vehicles and equipment.

The simulation used for this study accomplishes brigade movement with LCACs primarily transporting equipment and heavy vehicles. Rotary wing aircraft are the primary source of MEB sustainment. The JHSV transports Army supplies and light equipment exclusively; there is no need to burden a pier with heavy vehicles from a JHSV.

---

[22] In Appendix A, we assume alternatively that the distance from the sea base to the MEB is the same as the distance from the sea base to the SPOD, such as when the Marines are operating in the vicinity of the SPOD.

Because assets transporting Army brigades are also used to sustain Marines ashore, the nature of the Marine Corps force ashore and its level of combat will affect the movement of the Army brigade. The analysis treats an SBME engaged in heavy combat, an SBE engaged in sustained combat, and (as a worst case) an SBE engaged in heavy combat.[23] For Army forces transloading ashore via the MPF(F), it was assumed that the Army personnel had received sufficient training that they could move their vehicles safely from LMSRs via ramps onto the MLP and LCACs. Additionally, the Army's LMSRs were assumed to be loaded in a way that would facilitate selective offloading of vehicles and equipment.

Movement Without a JHSV

The analysis begins with the SBCT movement. Results (shown in Figure 3.11) suggest that an SBCT could be inserted using sea base assets in about three to five days for sea base distances of 25 to 50 NM to the SPOD. HBCT movement through the sea base (shown in Figure 3.12) would take about a day longer than SBCT movement. Results for the SBE in sustained combat (not shown) are very similar to those for the SBME in heavy combat.[24] The difference between the best case (the SBME in heavy combat) and the worst case (the SBE in heavy combat) is less than a day, and slight differences are seen between the cases of an SBE in sustained combat and an SBME in heavy combat. In operational terms, with over 20,000 tons of supplies and equipment passing through the sea base in several days for HBCT movement, the difference of several hundred tons a day in sustainment is modest.

We conclude that, in the context of moving an HBCT through the sea base as quickly as possible, SBME or SBE level of battle has a modest influence on movement time.

Our simulation used relatively few MV-22 sorties to transport Army personnel ashore in Scenario C. LCACs transported most Army

---

[23] Consumption rates for these cases are described in Appendix D.

[24] This finding is motivated in Appendix D.

Figure 3.11
SBCT Movement Using Aircraft and LCACs

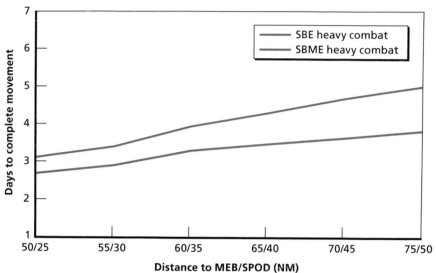

personnel (24 at a time[25]) as vehicles and equipment were transported ashore. As noted earlier, this simplified linking up Army troops with their vehicles and equipment ashore is a significant side benefit of this practice. Recall that the operation of the simulation reflected certain preferences, such as the movement of vehicles by LCACs and the movement of personnel by MV-22 aircraft. Further analysis showed that there is, in fact, no need to transport Army personnel by MV-22 aircraft in this scenario; LCACs could transport all Army personnel with their vehicles and equipment.[26]

---

[25] Marine Corps Combat Development Command (MCCDC), MSTP Center, *MAGTF Planner's Reference Manual*, Quantico, Va.: MSTP Pamphlet 5-0.3, 2006c.

[26] In our simulation, LCACs generated over 200 sorties in transporting Army SBCT vehicles and equipment ashore. At 24 passengers per load, this equates to a potential to transport about 5,000 passengers by LCACs. The Stryker brigade, the Army unit with the most personnel for this analysis, has 3,929 troops—significantly fewer than could be transported by LCACs. This finding also suggests that fewer troops could be transported by LCAC when passenger weight is an issue.

Figure 3.12
HBCT Movement Using Aircraft and LCACs

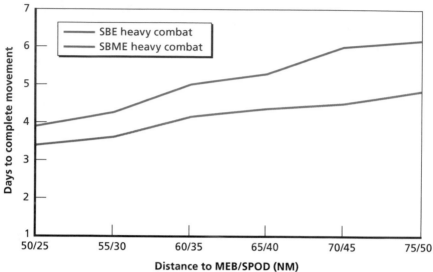

Movement with a JHSV

We assume now that one JHSV is available to assist LCACs and MPF(F) aircraft in moving an SBCT or an HBCT through a sea base, and that prevailing sea states allow the JHSV to transload personnel, supplies, and equipment at the MLP.[27] If the sea state and other factors permit its use, a single JHSV nearly halves SBCT movement time through the sea base (Figure 3.13) and roughly halves HBCT movement time through the sea base (Figure 3.14). Both of these figures indicate reductions in the effect of Marine Corps sustainment levels.

---

[27] A draft JHSV performance specification, Naval Sea Systems Command, SEA 05, *Joint High Speed Vessel (JHSV) Performance Specification (Draft)*, Working Paper, April 2007, circulated at the time of this analysis directs that the JHSV ramp system shall be designed, at a minimum, to support the loads associated with the M1A2 Abrams MBT weighing 80 short tons and the point loads generated by a fully loaded M1070 Military Truck and Trailer with a per axle weight of 32 short tons. It further specifies that the ramp shall be operable with these loads through Sea State 1 with the discharge end supported afloat. The implications of this draft requirement for lighter vehicles are unclear.

Figure 3.13
SBCT Movement in Scenario C, with and without a JHSV

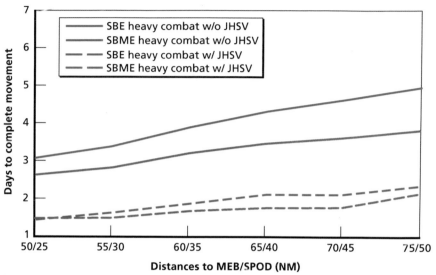

The repeated finding that a JHSV could roughly halve movement time is explained by three simple observations. First, LCACs are used here primarily to transport Army vehicles and equipment.[28] Second, MEB sustainment falls naturally to CH-53K and MV-22 aircraft in this scenario. Finally, the JHSV has load capacity (measured in square feet or tons) comparable to all 17 LCACs combined and is slightly faster than an LCAC; JHSV throughput roughly matches the combined throughput capacity of all the LCACs. The throughput of the JHSV and the LCACs combined is about twice that of the LCACs alone, resulting in about-halved Army brigade throughput time.

---

[28] The JHSV used in this analysis was taken from a recent JHSV AoA conducted by RAND (John F. Schank, Irv Blickstein, Mark V. Arena, Robert W. Button, Jessie Riposo, James Dryden, John Birkler, Raj Raman, Aimee Bower, Jerry M. Sollinger, and Gordon T. Lee, *Joint High-Speed Vessel Analysis of Alternatives*, Santa Monica, Calif.: RAND Corporation, 2006, not available to the general public). Of a number of candidates considered, it is at the median in capacity. It is also broadly consistent with the draft performance specifications for the JHSV.

Figure 3.14
HBCT Movement in Scenario C, with and without a JHSV

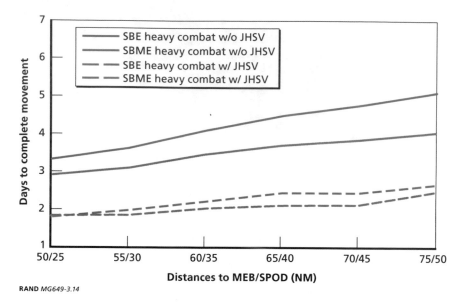

JHSVs are expected to carry several hundred passengers for short periods. A single JHSV would then add significant troop movement capacity to this operation. A corollary to this is the observation that a JHSV would further reduce the need for MV-22s as troop transports, freeing them for other missions.

Increasing the Ratio of CH-53K to MV-22 Aircraft

Our analysis of Scenario A demonstrated that increasing the ratio of CH-53K to MV-22 aircraft would improve sustainment performance. However, would this improvement come at the expense of Army BCT mobility? Our analysis indicates that changing the mix of MPF(F) aircraft as before would modestly reduce the time required to move an SBCT or an HBCT through a sea base (see Figure 3.15).

The explanation for this (possibly counterintuitive) finding is that the original aircraft mix (16 CH-53K and 34 MV-22 aircraft) is less efficient at sustainment than our changed mix (34 CH-53K and

Figure 3.15
SBCT Movement in Scenario C, with Differing Aircraft Mixes

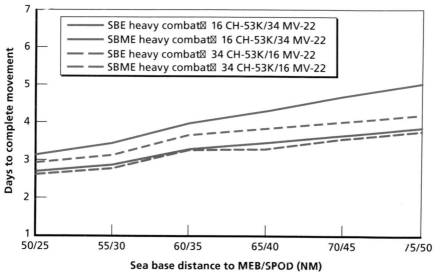

RAND MG649-3.15

16 MV-22 aircraft). Consequently, with MEB sustainment having higher priority than Army BCT movement, relatively few aircraft can be spared for BCT movement. In addition, as previously noted, the MV-22 is best suited for personnel movements (especially at long distances). However, there is little demand for airborne movement of personnel in this scenario. This finding reflects the LCAC's sidecar capacity for 24 personnel (two more personnel than the MV-22 can carry), who can be carried along with regular loads. In examining JSLM output, we found that less than 10 percent of MV-22 sorties were used for Army personnel movement. For both SBCT and HBCT movement, LCACs generated over 200 sorties to transport Army vehicles and equipment ashore. At 24 passengers per load, this history implies a potential for LCACs to transport over 5,000 personnel. The SBCT has 3,929 personnel, and the heavy brigade has 3,114 personnel. With an excess capacity for personnel movement using LCACs alone, it is clear why there was little demand for MV-22 sorties to transport personnel.

We turn finally to the Army's desire to operate some of its helicopters from the sea. In the context of flowing an Army brigade through the sea base, we consider the possibility that Army helicopters might be pre-positioned on the sea base for some time.

Army Helicopters on the Sea Base

Two operational concepts apply here: the use of sea base ships as "lily pads" for occasional transfer operations between the sea base and the shore, and the location of tens of Army helicopters on the sea base.

Under the former concept, Army CH-47 helicopters might periodically use open operational spots on an MPF(F) LHA(R) or the LHD. Under the latter concept, embarked aircraft would have to be removed from those ships to make space for Army helicopters. In either concept, Army helicopter operations would interfere to some degree with shipboard sustainment operations.

In the lily-pad concept, Army helicopters would spend little time at sea, so preparing them for operations in a maritime environment (such as providing corrosion protection) would probably not be an issue. The question then is, To what extent would such operations interfere with sustainment operations? We found that operational spots would not be available at the outset of the flight window as the LHA(R)/ LHD launched CH-53/MV-22 aircraft and recovered those aircraft it had launched initially. In particular, MPF(F) helicopters unfolding on operational spots could not readily be moved to accommodate Army helicopters in this period. However, the aircraft would not operate in dense waves after a few hours; Army aircraft could then land and take off with little or no interference with sustainment operations. Even in the worst case, there would normally be flight spots available for Army-helicopter takeoff and landing. As a worst case, we assumed that one or more flight spots were assigned for the duration to Army helicopters.

The difficulties inherent in operating U.S. Army and Air Force helicopters from Navy ships are well known; Army and Air Force helicopters operated from Navy ships during contingency operations in Grenada, Panama, Somalia, and Haiti. Moreover, U.S. Army Field

Manual FM 1-564, *Shipboard Operations*,[29] describes the tactics, techniques, and procedures for use by Army aviation units during operations from Navy and Coast Guard ships. This publication addresses the problems of operating Army helicopters from Navy ships, including

- lack of rotor brakes. Army helicopters lack rotor brakes to rapidly slow blades when the helicopter engine is shut off; without rotor brakes, helicopter blades can windmill for several minutes, slowing shipboard operations and creating hazardous conditions for shipboard personnel. The absence of rotor brakes also poses a threat to the helicopter. FM 1-564 states,

    > The ship must be kept on a steady course and speed during rotor engagement or disengagement, engine start and shutdown for aircraft without rotor brakes, taxiing, and launch or recovery operations. Deck tilt, centrifugal force, or rapidly changing wind direction or velocity aerodynamically affects the controllability of the aircraft and may cause rollover.

- blade folding. Army helicopters must be modified for a folding capability needed to operate on Navy ships. Aircraft modified with a blade folding capability must deploy with the proper blade folding kit to allow movement into hangars. When these aircraft are positioned on the flight deck, they are vulnerable to damage when the blades flap in the wind.

There are well-known fuel flashpoint issues with Army aviation fuel. Moreover, Army aircraft are not manufactured to the anticorrosion standards of Navy aircraft and are prone to corrosion; experience has shown that unprotected major aircraft components can lose an estimated 25 to 30 percent of their useful life through saltwater corrosion. Units should obtain an anticorrosion compound for their aircraft before embarkation. Freshwater washes may not be conducted as frequently as desired. Army pilots must be qualified with overwater training, daytime and nighttime landings, and for any logistics over

---

[29] Headquarters, Department of the Army, Washington, D.C., June 1997.

the shore or vertical replenishments. Unlike the rotor brake and blade folding problems, these other issues have clearly identified remedial procedures.

We assume that the above issues can be resolved and look at the feasibility of positioning Army helicopters on the sea base—landing them on MPF(F) ships, locating them on spots normally occupied by aircraft, and flying them off at the appropriate time (again, assuming that MV-22 aircraft have been moved ashore to make space for CH-47F aircraft). Of the 34 operational MV-22 aircraft earmarked for sustainment, how many could be moved off the MPF(F) ships without breaking sustainment capability? As above, this analysis is conducted in Appendix E under Scenario A, which is more stressing than Scenario B on aircraft sustainment operations.

The CH-47F is significantly smaller than the MV-22 when both are folded.[30] A size comparison suggests that, with 34 MV-22 aircraft removed (presumably ashore) from the MPF(F) ships, over 40 CH-47F aircraft could be located on the MPF(F) ships.

Absent LCACs and MV-22s (i.e., using only CH-53K helicopters), MPF(F) ships appear incapable of sustaining an SBME and an airborne combat brigade in heavy combat. A robust sustainment capability is seen when LCACs also are used to sustain the SBME. We conclude that, with corrosion and rotor issues addressed and with all MV-22 aircraft relocated ashore or elsewhere, over 40 CH-47F helicopters could be positioned on the MPF(F) ship while it sustained an SBME and an airborne brigade.

---

[30] The CH-47F is 50 feet long when folded, whereas the MV-22 is 63 feet long when folded. The CH-47F is 12 feet, 5 inches wide, and the MV-22 is 18 feet, 5 inches wide.

CHAPTER FOUR

Conclusions

Overall Findings

Simultaneous sustainment of brigade-level Army and Marine Corps ground elements using planned MPF(F) components is feasible. Issues of sustainment under unfavorable conditions, such as in high sea states with degraded ship-to-ship movement, can be addressed in part using the metric of relative sustainment capacity. Overcapacity (under favorable conditions) is needed for adequate capacity under unfavorable conditions.

Additionally, with overcapacity, sea base assets (notably, MV-22 aircraft) can be released to ground forces under favorable conditions.

We identified the following distinct approaches to increasing sustainment capacity, along with the following findings:

- Reducing distances from the large-deck MPF(F) and MLP ships to supported ground elements or seaports of debarkation. Reducing sustainment distances from the planned distance of 110 NM is the most effective means of increasing sustainment capacity. Threat conditions can, of course, limit this option, necessitating others.
- Adding LCAC surface connectors to CH-53 and MV-22 aircraft in sustainment. The addition of LCACs operating 16 hours a day more than doubles sustainment throughput.
- Increasing the ratio of CH-53K to MV-22 aircraft. Increasing the ratio of CH-53K to MV-22 aircraft can have benefits similar to those from adding LCACs as sustainment assets.

- Reducing sustainment requirements. Reducing demand for external sustainment, such as that realized by eliminating demand by ground elements for bulk water, can significantly improve the ability to sustain ground elements. In particular, it can extend the maximum distance from the sea base at which ground elements can be sustained.

The following approaches to reducing times for Army ground element movement were identified, along with the following findings:

- Increasing the ratio of CH-53K to MV-22 aircraft. A modest reduction in movement time can be achieved by increasing the ratio of CH-53K to MV-22 aircraft. Put another way, it would enhance sustainment performance significantly without increasing movement time.
- Adding JHSVs to LCACs as surface connectors. A single JHSV about equals the combined lift capacities of LCACs from the sea base. In this light, adding a JHSV to LCACs roughly doubles surface connector movement capacity.

Sustainment Results

Our analysis of SBME sustainment indicates that an SBME can be sustained, with some difficulty, at a range of up to 110 NM from the sea base, using only CH-53K and MV-22 aircraft. An SBE can be sustained similarly at ranges up to about 70 NM from the sea base. By reducing the distance from the sea base to the MEB to about 80 NM, an SBME and an Army airborne brigade can be sustained simultaneously using only aircraft.[1] An SBE and an airborne brigade can likewise be sustained when the MEB distance is up to about 40 NM from the sea base.

---

[1] This analysis assumes that Army brigades are 50 NM farther away than the MEB from the sea base. With the MEB 80 NM from the sea base, the airborne brigade would be 130 NM from the sea base.

Using LCACs to augment sea base aircraft in sustainment has substantial benefits, particularly when LCACs contribute to both Marine Corps and Army ground element sustainment. Under conditions in which LCACs can contribute only to MEB sustainment, the limitations of airborne sustainment to Army ground elements determine the feasibility of joint sustainment. Here, a mix of sea base aircraft richer in CH-53K aircraft is less limited than the planned aircraft mix, enabling joint sustainment at greater distances. As noted earlier, we assumed that the forward movement of supplies delivered to the shoreline by the LCACs would be tactically feasible and that sufficient Army and Marine Corps trucks would be available to conduct the movement.

Reducing sustainment demand (by, for example, eliminating demand for bulk water from the sea base) is particularly helpful when sustainment capacity is marginal. For example, it would increase by about 25 NM the distance at which an SBE and an airborne brigade can be sustained using only aircraft. Note that this study assumes that supplies for Army units would not come from Marine Corps stocks aboard the MPF(F) ships. Rather, it was assumed that other shipping would be available to bring Army supplies into the operational area for transfer ashore by sea-based aircraft and LCACs. The details of how that forward movement of supplies would be accomplished was beyond the purview of this study, but the issue clearly merits additional analysis of how sensitive the onward movement of supplies would be to enemy threats and the number of trucks that might be available.

Movement Results

An Army Stryker or heavy brigade can be transloaded at sea and moved ashore from the sea base in three to six days (depending on the distance off shore), using MPF(F) assets also sustaining a MEB—a new capability for the Army.

If a single JHSV can augment the LCACs, it will roughly halve the time required to transport an Army brigade ashore. This finding reflects the observation that, when operable, the throughput capacity of a single JHSV about matches the combined throughput of MLP

LCACs. There are, however, issues of JHSV operability in this role in even moderate sea states (Sea State 2 or higher), as well as the need for a small port where the JHSV can offload.

Other Findings

- The CH-53K is better suited than the MV-22 for sustainment; with external loads, the MV-22 loses its speed advantage on ingress and the CH-53K carries at least twice the load of the MV-22. CH-53K helicopters are especially valuable under conditions of heavy sustainment demand or long sustainment distances.
- The Sea Base concept is not consistent with, and in some sense conflicts with, the Army's desire to deploy directly to a port via High-Speed Ships. The Army has not developed doctrine and has not funded systems for operating with sea bases. However, our Scenario B analysis illustrates that, once ashore, an Army brigade could in many situations be sustained by a sea base if it (1) moved away from its port of debarkation or (2) that port became unavailable for sustainment as a result of enemy action.
- To capitalize on the potential of the sea base, Army shipping should be configured for "selective offload" or "combat loading" rather than "dense pack." The interface between Army prepositioning ships and the MLP is a potential bottleneck in moving Army forces. Thought should be given to an MLP loading system built into the MLP to avoid such bottlenecks. Integrating such a loading system into the MLP might be less expensive in net than integrating it into Army and Navy pre-positioning ships. It might also hasten joint interoperability.
- MPF(F) ships can provide temporary deck space (1–2 deck spots per "big deck") for a limited number of Army helicopters without significant loss of throughput capacity. There is not sufficient space on the MPF(F) to base a significant number of Army aircraft as long as a large number of Marine Corps MV-22s and

CH-53Ks are based on the MPF(F). Space for Army aircraft could be created temporarily by moving MV-22 aircraft ashore, but several problems would remain, including rotor issues (braking and folding), corrosion, and maintenance.

Additional Cases

This appendix presents cases omitted for brevity in the main body of this monograph. Our purpose here is to examine additional cases of interest, to illustrate the generality of findings in the main body and to better illuminate dependencies. As in Chapter Three, cases here are organized by scenario.

Scenario A—Army Forces Arrive Inland

MLPs 50 NM from the SPOD

MLPs were assumed in Chapters Two and Three to operate 25 NM from the SPOD when LCACs were used in sustainment. A minor (10 to 20 percent) reduction in the benefits of adding LCACs in sustainment is seen when MLPs operate 50 NM from the SPOD. We infer, then, that in the range of 25 to 50 NM, this distance is not critical to our findings.

Other than the assumption that MLPs are 50 NM from the SPOD, all other conditions here are identical to those used to generate Figure 3.4. Results shown in Figure A.1 can be compared with those shown in Figure 3.4. The loss of sustainment capacity, for both the SBE and the SBME in this comparison, is uniformly 10 to 20 percent, which is considered minor. The breakpoint for sustainment to the airborne brigade remains unchanged: changing the distance from the MLPs to the SPOD does not change CH-53K and MV-22 sustainment performance, and the breakpoint depends solely on the ability of these

Figure A.1
VTOL Plus LCAC Sustainment in Scenario A, with MLPs 50 NM from SPOD

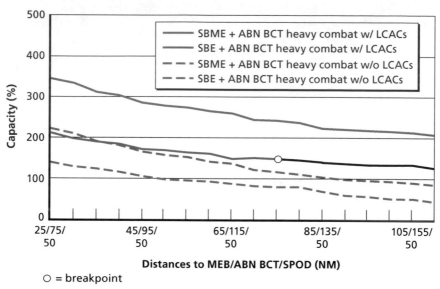

O = breakpoint

aircraft to meet Army BCT needs. As for the LCACs, their number of sorties was reduced by about 25 percent.[1] With aircraft performance unchanged and LCAC throughput decreased by 25 percent, the net effect of increasing the distance to the SPOD from 25 to 50 NM was reduction of overall throughput by 10 to 20 percent.

LCACs Operating 12 Hours per Day

Given a single LCAC crew for each LCAC, crew fatigue limits LCACs to 12 hours or less of operation per day. Marine Corps planners commonly assume that LCAC crews are adequate for 16 hours of operation per day,[2] and this assumption was used in Chapters Two and Three. We now assume that LCACs are limited by crew fatigue to 12 hours

---

[1] LCAC load and offload times per sortie were unchanged. Hence doubling transit times did not double sortie durations or halve the number of sorties.

[2] MCCDC, MSTP Center, *MAGTF Planner's Reference Manual*, Quantico, Va.: MSTP Pamphlet 5-0.3, 2006c, p. 35.

of operation per day. As above, a minor (10 to 20 percent) reduction in the benefits of adding LCACs in sustainment is seen when LCACs operate no more than 12 hours a day. The ability to operate LCACs 16 hours per day is helpful but not critical to our findings.

All conditions for this analysis (other than the assumption that LCACs and MLPs operate no more than 12 hours per day) are identical to those used to generate Figure 3.4; results shown in Figure A.2 can be compared with those shown in Figure 3.4. The loss of sustainment capacity seen in this comparison, for both the SBE and the SBME, is uniformly 10 to 20 percent, again considered minor. The number of LCAC sorties per day was reduced by about 25 percent, and the number of aircraft sorties available for Army BCT sustainment was reduced slightly. The net effect of limiting LCACs to not more than

Figure A.2
VTOL Plus LCAC Sustainment in Scenario A, with LCACs Limited to 12 Hours of Operation per Day

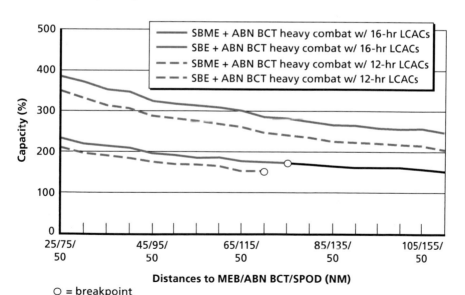

O = breakpoint

RAND *MG649-A.2*

12 hours of operation per day was reduction of overall throughput by 10 to 20 percent.

SBE in Sustained Operations

MEB results were presented in Chapter Three for an SBE or an SBME in heavy combat, generally in combination with an Army brigade. We now add the results for an SBE in sustained combat with the results for it falling between those for an SBE or an SBME in heavy combat (more closely resembling those for an SBE in heavy combat).

We begin with air-only sustainment of an SBE in sustained operations, along with an airborne brigade, as in Figure 3.3. Here, as shown in Figure A.3, the SBE in sustained operations falls between the SBME in heavy combat and the SBE in heavy combat.

Further illustration is provided in Figure A.4, which shows the result of adding LCACs to MPF(F) aircraft in sustainment. More

Figure A.3
VTOL-Only Sustainment of a MEB and an Army Airborne Brigade in Scenario A Is Marginal

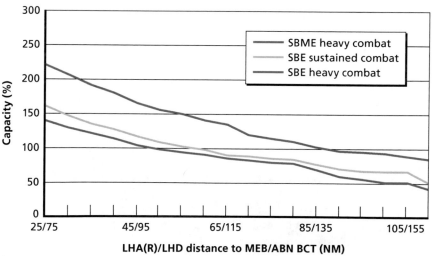

RAND MG649-A.3

Figure A.4
VTOL Plus LCAC Sustainment in Scenario A Is More Robust

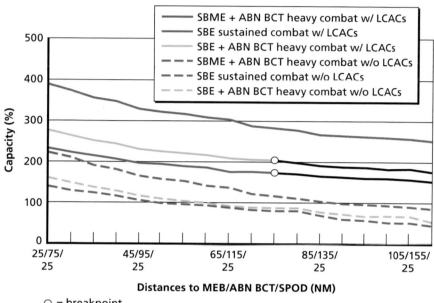

○ = breakpoint

robust capability is again seen, with results for the SBE in sustained operations falling between the SBE in heavy combat and the SBME in heavy combat. As noted in the preceding section, breakpoints depend solely on the ability of MPF(F) aircraft to meet Army BCT needs, so they are unchanged.

We conclude generally that sustainment results for an SBE in sustained combat fall between those for the SBE in heavy combat and the SBME in heavy combat.

Sustainment with a Reduced Number of MV-22s

One conclusion of this study is that there is insufficient space on the MPF(F) to base a significant number of Army aircraft as long as large numbers of Marine Corps MV-22s and CH-53Ks are based on the MPF(F). Here, we find that in SBME sustainment before the arrival of an Army ground element, all MV-22s could be put ashore to make room for Army CH-47 helicopters either by using LCACs for sustainment or by reducing the distance from the large deck ships to the SBME.

Relative sustainment capacity for an SBME in heavy combat is shown in Figure A.5 for CH-53K helicopters alone and for CH-53K helicopters working with LCACs. Figure A.3 results with MV-22 aircraft are also shown here for reference.

As a secondary finding, Figure A.5 also shows that MV-22 aircraft contribute relatively little to sustainment for large distances (i.e., distances approaching 110 NM).

Figure A.5
SBME Sustainment in Scenario A, Without MV-22 Aircraft

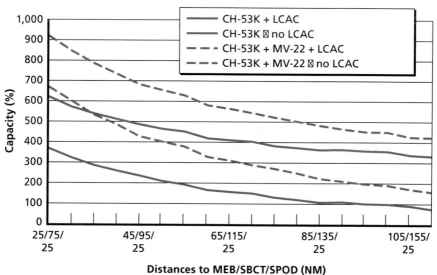

Sustainment with Varying Numbers of Operational Flight Spots

Here, we address an issue raised in Chapters Two and Three and considered in the MPF(F) CDD analysis: the effect of dedicating operating spots on the three large flight decks for operations other than sustainment.[3]

Results of dedicating additional flight spots, with and without LCACs, are shown in Figure A.6 for an SBME and an airborne brigade in heavy combat. Results with LCACs are shown using sold lines; results without LCACs are shown using dashed lines. With or without LCACs, the effect of dedicating one or two additional operating spots on each of the three large decks is minor.

Figure A.6
Dedicating Additional Operating Spots for SBME, Airborne BCT in Heavy Combat

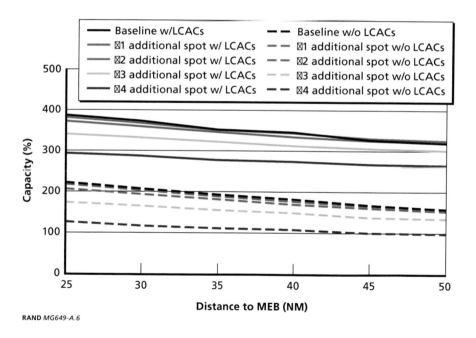

---

[3] A general assumption of this study is that one operational spot on each of the three large decks will be dedicated to MV-22 combat search-and-rescue aircraft.

Scenario B—Army Forces Enter the Area of Operations Directly

MLPs 50 NM from the SPOD

Paralleling the previous section, how much would the benefits of using LCACs be reduced in Scenario B if the MLP ships operated 50 NM from the SPOD? Capacity is seen to be reduced by 10 to 20 percent, again considered minor.

Other than this increase in distance, conditions here match those used for Figures 3.9 and 3.10, which can be compared against the results shown in Figures A.7 and A.8.

In both Scenarios A and B, the effect of increasing the MLP distance to the SPOD when LCACs are used is in the range of 10 to 20 percent. We conclude generally that increasing the distance from the MLPs to the SPOD from 25 NM to 50 NM has only a minor effect on relative capacity.

Figure A.7
MEB Plus SBCT Sustainment in Scenario B, with MLPs 50 NM from SPOD

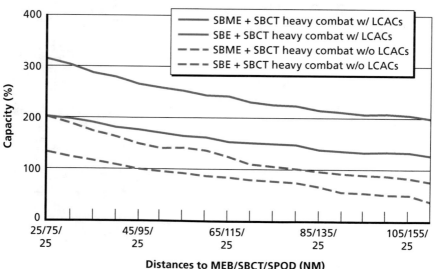

RAND *MG649-A.7*

Figure A.8
MEB Plus HBCT Sustainment in Scenario B, with MLPs 50 NM from SPOD

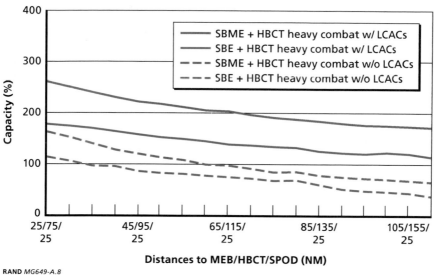

RAND *MG649-A.8*

Increasing the Ratio of CH-53K to MV-22 Aircraft

Chapters Two and Three considered the implications of using a richer mix of CH-53K to MV-22 aircraft only in the context of Scenario A. The topic is taken up again here with the finding that reversing the ratio of CH-53K to MV-22 aircraft has a significant advantage in sustainment.

We consider four cases: sustainment for an SBME plus an SBCT; for an SBE plus an SBCT; for an SBME plus an HBCT; and for an SBE plus an HBCT. Only airborne sustainment cases (i.e., cases without LCACs) were considered to bound the analysis. For all cases, increasing the ratio of CH-53K to MV-22 aircraft significantly increases both the robustness of sustainment and the maximum distances from which sustainment is possible (Figures A.9 and A.10).

Figure A.9
SBCT Sustainment in Scenario B, with Altered Aircraft Mix

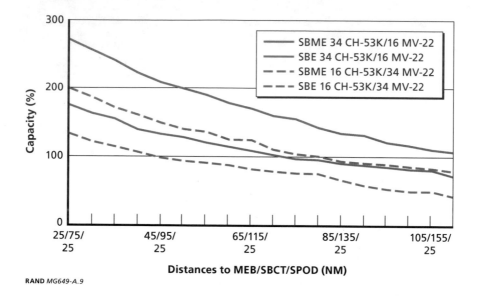

RAND *MG649-A.9*

Reducing Sustainment Demand

The implications of reducing sustainment demand, exemplified by eliminating the need for bulk water from the sea base, were considered earlier in this monograph in the context of Scenario A. Significant sustainment benefits for Stryker and heavy brigades in heavy combat, similar to those seen before, are seen in Figures A.11 and A.12, respectively.

Figure A.10
HBCT Sustainment in Scenario B, with Altered Aircraft Mix

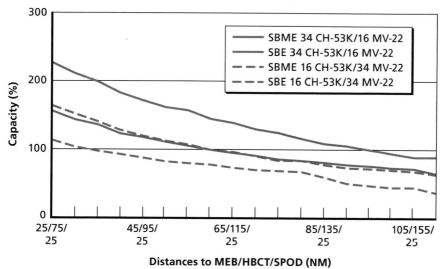

RAND *MG649-A.10*

Figure A.11
SBCT Sustainment in Scenario B, with and without Bulk Water

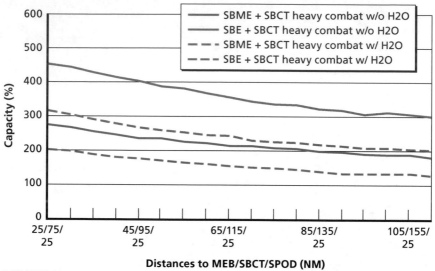

Figure A.12
HBCT Sustainment in Scenario B, with and without Bulk Water

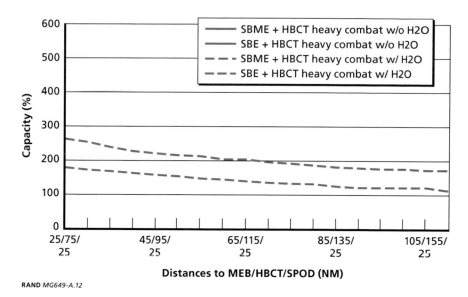

RAND *MG649-A.12*

Scenario C—Army Forces Enter the Area of Operations via the Sea Base

Previously, the MEB was assumed to operate inland, and sustainment to it was delivered 25 NM farther than the Army objective area. In other words, the Army brigade is inserted from distances of 25 to 50 NM, whereas the MEB is sustained from distances of 50 to 75 NM. Now, we assume that the distance from the sea base to the MEB is the same as the distance from the sea base to the seaport of debarkation, such as when the Marines are operating near the SPOD. This difference (seen in Figures A.13 and A.14) can reduce the time to complete movement of an SBCT or an HBCT by over a day, depending on distances to the SPOD.

Figure A.13
SBCT Movement for Differing MEB Locations

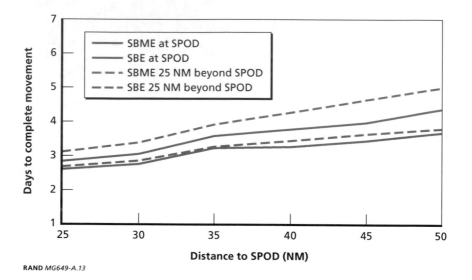

RAND *MG649-A.13*

Figure A.14
HBCT Movement for Differing MEB Locations

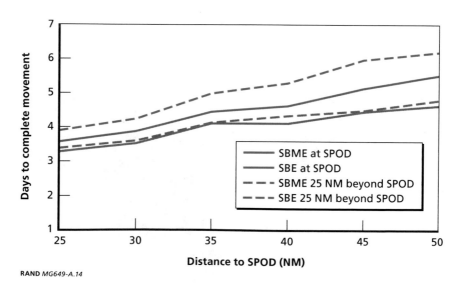

RAND *MG649-A.14*

Maritime Pre-positioning Force (Future) Description

At the time of this analysis, the planned Maritime Pre-positioning Force (Future) (MPF(F)) squadron will comprise two LHA Replacement (LHA(R)) large-deck amphibious ships; one modified LHD large-deck amphibious ship; three modified Lewis and Clark (T-AKE) cargo ships; three modified Large, Medium-Speed, Roll-on/Roll-off (LMSR) sealift ships; three Mobile Landing Platform (MLP) Landing Craft Air Cushion (LCAC) transport ships; and two legacy "dense pack" MPF ships taken from existing squadrons.

This appendix describes these ships, except for the existing MPF ships, which are not relevant to this study.[1]

LHA(R) and LHD

The notional LHA(R) Flight 0 large-deck amphibious ship will be a modified version of the LHD-8 amphibious assault ship. Designated LHA-6, it is notable for its lack of a well deck, which means that it cannot operate LCACs or landing craft, utility (LCU) ships. It will have nine Aviation Landing Spots, six on the port side. An MPF(F) LHA(R) is distinguished from an Expeditionary Strike Group (ESG) LHA(R) by its simplified command and control system and lack of active defense systems. It will be able to operate three LCAC-

---

[1] The Programs and Resources Branch of the Marine Corps updates MPF(F) program information annually.

equivalent connectors, and it will have nine Aviation Landing Spots, seven on the port side. Future LHA(R)s will also be developed in ESG and MPF(F) versions. The MPF(F) LHD will be a decommissioned LHD from the fleet, modified for MPF(F).

MPF(F) LHA(R) and LHD vessels are to collectively carry a 2015 MEB Air Combat Element to include 48 MV-22, 20 CH-53K, and 18 AH-1 helicopters. Each aviation ship is to carry two SH-60 helicopters.[2] Both the LHA(R) and the LHD will store 400,000 gallons of water and produce 200,000 gallons of water per day.

A current LHD, the USS *Bataan* (LHD-5), is shown in Figure B.1, with MV-22 aircraft spotted.

T-AKE Cargo Ships

The T-AKE is a new Combat Logistics Force (CLF) Underway Replenishment Naval vessel, originally known as the Auxiliary Dry Cargo Carrier (ADC(X)). It has two multipurpose cargo holds, capable of selective offload, for dry stores and/or ammunition. It has additional holds for freeze, chill, and/or dry stores, and three specialty and spare parts cargo holds. Its cargo capacity for dry cargo/ammunition is approximately 1,100,000 square feet. Fuel capacity is 1,300,000 gallons. Water capacity is 52,800 gallons, and it has a capacity to produce 28,000 gallons of water per day.

The T-AKE has a single vertical replenishment (VERTREP) station. Its design speed is 20 knots. The lead ship of the class, T-AKE-1, the USNS *Lewis and Clark*, was delivered to the U.S. Navy in June 2006. It is shown in Figure B.2.

---

[2]   The Secretary of the Navy approved this squadron in May 2005.

Figure B.1
LHD-5, USS *Bataan*

SOURCE: U.S. Navy, V-22 program Web site.

RAND *MG649-B.1*

LMSR Cargo Ships

The MPF(F) LMSR will have about 202,000 square feet of cargo space and two or four aircraft operating spots, and it will berth about 850 personnel. Its design speed is 20 knots. It will store 33,500 gallons of water, and it will have the capacity to produce 24,000 gallons of water per day.

Figure B.3 illustrates an MPF(F) LMSR alongside an MLP. Note the ramp on the LMSR that is lowered between the two ships.

Figure B.2
T-AKE-1, USNS *Lewis and Clark*

SOURCE: U.S. Navy, Military Sealift Command Ship Inventory.
RAND *MG649-B.2*

Mobile Landing Platform

The Mobile Landing Platform will be a clean sheet design, leveraging existing float-on/float-off technology. It is to carry six LCAC-equivalent connectors and one Brigade Landing Team (BLT) of equipment, and it will have accommodations for 1,458 personnel. It will have one aircraft landing spot. Its design speed will be about 20 knots. Planned fuel capacity will be about 1,200,000 gallons. It is expected to carry 168,000 gallons of water; its water production capacity is still under consideration.

LCACs cannot operate 24 hours a day. As discussed in Appendix E, MLP operating days were matched to the LCAC operating day.

Figure B.3
MPF(F) LMSR Alongside an MLP

SOURCE: Office of the Chief of Naval Operations (N81).
RAND *MG649-B.3*

Figure B.4 illustrates an MLP transferring vehicles onto a notional JHSV while alongside an existing LMSR. Other mooring configurations are possible between the JHSV and the MLP.

Figure B.4
MLP Operations

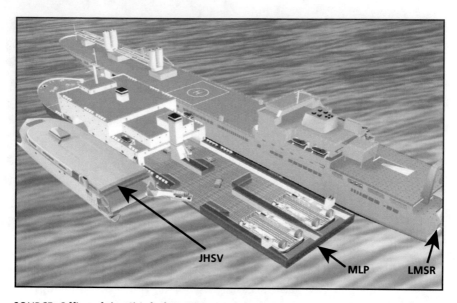

SOURCE: Office of the Chief of Naval Operations (N81).

Army and Marine Corps Ground Elements Evaluated

Army Elements

The Army has three types of brigade combat teams today: the Infantry Brigade Combat Team (IBCT), the Stryker Brigade Combat Team (SBCT), and a heavy Brigade Combat Team (HBCT). The airborne brigade analyzed in this study is a form of Infantry Brigade. This section describes the personnel and equipment of the IBCT, SBCT, and HBCT, respectively.

Infantry Brigade Combat Team

The Infantry Brigade Combat Team, as recently described by the Army,[1] is illustrated in Figure C.1.

The IBCT has about 3,600 soldiers. It has 592 vehicles with a total weight of 2,105 short tons (ST). The total weight of the IBCT is 2,360 ST. Major equipment items[2] in the IBCT are shown in Table C.1.

---

[1] Secretary of the Army, "The Army Modular Force 2004–2020," briefing, no date.

[2] Major equipment lists do not include such items as water or flatbed trailers, dump trucks, loaders, electrical generator sets, or unit equipment.

Figure C.1
Design of the Army IBCT

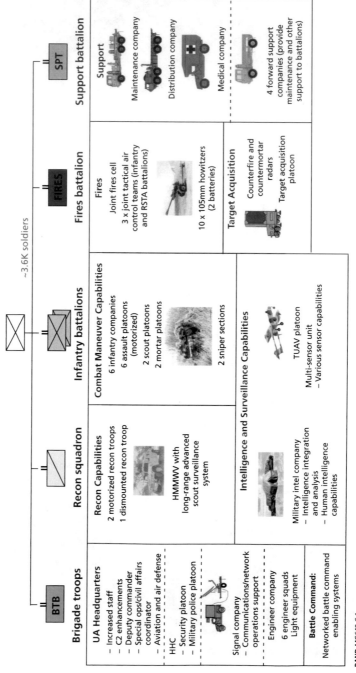

RAND *MG649-C.1*

Table C.1
Major Equipment in the Infantry Brigade Combat Team

Major Items of Equipment	Items
Lightweight Towed 105mm Howitzer	8
Mortar, 120mm	4
High-Mobility Multipurpose Wheeled Vehicle (HMMWV)[a]	75
Stryker Vehicle (M707)	7
Heavy Truck w/ Light Crane (M977/M977A2/M997A2R1)	16
Medium Tactical Vehicle, Cargo (M1084/M1089/M1083)	25
Light Medium Tactical Vehicle, Cargo (M1078)	13
Utility Truck (M1026A1/M966A1/M1113/M1025A2/M998A1/M1097A2)	263

SOURCE: Combined Arms Support Command, Fort Lee, Va.

[a] All HMMWVs, including ambulances.

Stryker Brigade Combat Team

The Stryker Brigade Combat Team, as described by the Army, is illustrated in Figure C.2.

The SBCT has about 3,900 soldiers. It has 1,551 vehicles that weigh a total of 13,567 ST. The total weight of the SBCT is 14,603 ST. Major equipment items in the SBCT are shown in Table C.2.

Heavy Brigade Combat Team

The heavy Brigade Combat Team is illustrated in Figure C.3.

Table C.2
Major Equipment in the Stryker Brigade Combat Team

Major Items of Equipment	Items
Stryker Vehicle[a]	302
Medium 155mm Howitzer (XM777)	18
Mortar Carrier, 120mm	30
High-Mobility Multipurpose Wheeled Vehicle (HMMWV)	7
Utility Truck (M1038A1/M1097A2/M1025A2/M998A1/M1113)	381
Cargo Truck (M1083A1/M1084A1)	158

[a] All Stryker vehicles.

Figure C.2
Design of the Army Stryker Brigade Combat Team

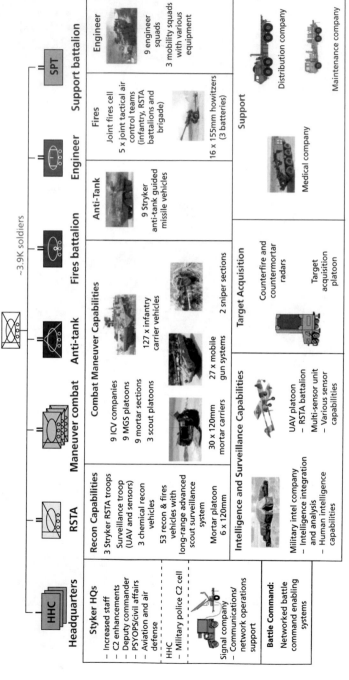

Figure C.3
Design of the Army Heavy Brigade Combat Team

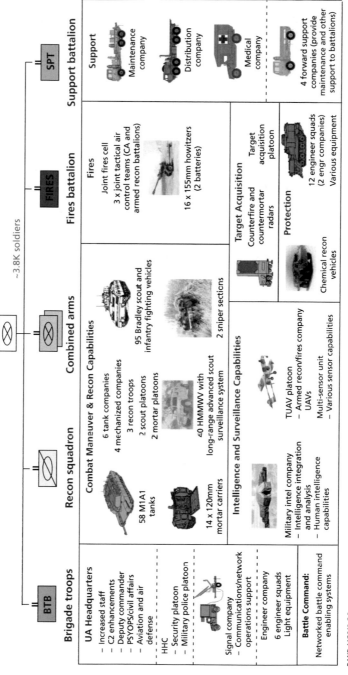

~3.8K soldiers

Brigade troops

BTB

UA Headquarters
– Increased staff
– C2 enhancements
– Deputy commander
– PSYOPS/civil affairs
– Aviation and air
  defense

HHC
– Security platoon
– Military police platoon

Signal company
– Communications/network
  operations support

Engineer company
6 engineer squads
Light equipment

Battle Command:
Networked battle command
enabling systems

Recon squadron

Combat Maneuver & Recon Capabilities

58 M1A1
tanks

14 x 120mm
mortar carriers

40 HMMWV with
long-range advanced scout
surveillance system

Intelligence and Surveillance Capabilities

Military intel company
– Intelligence integration
  and analysis
– Human intelligence
  capabilities

TUAV platoon
– Armed recon/fires company
  UAVs
Multi-sensor unit
– Various sensor capabilities

Combined arms

6 tank companies
4 mechanized companies
3 recon troops
2 scout platoons
2 mortar platoons

95 Bradley scout and
infantry fighting vehicles

2 sniper sections

Fires battalion

FIRES

Fires
Joint fires cell
3 x joint tactical air
control teams (CA and
armed recon battalions)

16 x 155mm howitzers
(2 batteries)

Target Acquisition
Counterfire and
countermortar
radars

Target
acquisition
platoon

Protection
Chemical recon
vehicles

12 engineer squads
(2 engr companies)
Various equipment

Support battalion

SPT

Support
Maintenance
company

Distribution
company

Medical
company

4 forward support
companies (provide
maintenance and other
support to battalions)

RAND MG649-C.3

The HBCT has about 3,800 soldiers. It has 1,770 vehicles with a total weight of 18,964 ST. The total weight of the HBCT is 20,202 ST. Major equipment items in the HBCT are shown in Table C.3.

MEB Elements

The Marine Corps developed the MPF(F) MEB, or 2015 MEB, for operation from MPF(F) ships.[3] This is the MEB evaluated here.

The 2015 MEB has 14,484 personnel, organized into a Sea Base Echelon (with 8,397 personnel), a Forward Base Echelon (with 1,907 personnel), and a Sustained Operations Ashore Echelon (with 4,180 personnel). The Sea Base Echelon (SBE) has a Sea Base Maneuver Element (SBME) with 4,989 personnel and a Sea Base Support Element (SBSE) with 3,408 personnel. The SBSE initially supports the SBME from the sea base; later, much of it will move ashore to better support the SBME. The Sustained Operations Ashore Echelon normally operates from the continental United States.

Table C.3
Major Equipment in the Heavy Brigade Combat Team

Major Items of Equipment	Items
Main Battle Tank (M1A2)	58
Bradley Fighting Vehicle (M2A3/M3A3)	109
Armored Personnel Carrier (M113A3)	43
Light Armored Vehicle (M1130/M1135)	4
Mortar Carrier, 120mm	14
High-Mobility Multipurpose Wheeled Vehicle (HMMWV)	45
Recovery Vehicle, Full Track Heavy (M88A1/M88A2)	23
Utility Truck (M1114/M1097A2/M1025A2/M1038A1/M998A1/M1113)	451
Cargo Truck (M1083A1/M1078A1/M985A2R1/M1074)	218
Stryker Vehicle (M707)	5

[3] Headquarters, United States Marine Corps, "Amphibious Requirements: USN and USMC Warfighter Talks," briefing, Washington, D.C., February 2, 2007.

Major equipment of the 2015 MEB include three squadrons of Joint Strike Fighters, a squadron of EA-18G Electronic Attack aircraft, and a squadron of light attack helicopters. These aircraft operate from the large deck MPF(F) ships (the two LHA(R)s and the LHD). Additional aircraft include 1.25 CH-53 helicopter squadrons (20 helicopters) and 4 MV-22 squadrons (48 aircraft). A KC-130 squadron operates ashore.

Other major equipment is shown in Table C.4.

Table C.4
Major Equipment in the MPF(F) MEB

Major Items of Equipment	Items
Expeditionary Fighting Vehicle (EFV)	106
Light Armored Vehicle (LAV)	54
Main Battle Tank (M1A1)	30
Lightweight 155mm Howitzer (LW155)	18
Expeditionary Fire Support System (EFSS)	8
High Mobility Artillery Rocket System	6
Joint Tactical Radio Set (JTRS)	202
High-Mobility Multipurpose Wheeled Vehicle (HMMWV)	717
Internally Transported Vehicle (ITV)	24
Medium Tactical Vehicle Replacement (MTVR)	236
Logistics Vehicle System (LVS)	131

Sustainment Requirements

Army and Marine Corps sustainment is treated, as in the Marine Corps Capabilities Development Document (CDD) analysis, as ammunition, dry stores, bulk water, and Petroleum, Oil, and Lubricants (POL) per day. Ammunition and dry stores are measured in short tons per day. Bulk water and POL are measured in gallons per day.

MEB consumption data for the cases examined in this study are shown in Table D.1. Number of personnel drives the consumption of water and dry stores; this explains the identical entries seen in the table below. The SBE has 8,397 personnel whereas the SBME has 4,989 personnel.

All Army consumption rates are for heavy combat. They are shown in Table D.2. For completeness, the airborne Brigade Combat Team (BCT) has 3,411 personnel; the Stryker BCT has 3,929 personnel, and the heavy BCT has 3,117 personnel.

Table D.1
Marine Corps Sustainment Requirements

	Ammunition (ST/day)	Dry Stores (ST/day)	Water (gal/day)	POL (gal/day)
Heavy combat				
SBME	250	33	31,621	78,637
SBE	339	51	60,264	230,227
Sustained				
SBE	75	51	60,264	176,305

Table D.2
Army Brigade Sustainment Requirements

	Ammunition (ST/day)	Dry Stores (ST/day)	Water (gal/day)	POL (gal/day)
Airborne BCT	7	51	36,907	24,640
Stryker BCT	41	71	42,512	29,719
Heavy BCT	62	84	33,726	83,971

The worst case for water consumption ashore is an SBE in combination with a Stryker BCT, which combine to consume about 103,000 gallons of water per day. This demand could be met by a single LHA(R) or LHD, or by three LMSRs and two T-AKEs, or by other combinations of MPF(F) ships.

A graphic comparison of these consumption rates is provided in Figure D.1. Bulk liquid consumption rates have been converted to short tons per day for clarity. The SBE in heavy combat, reflecting high consumption rates of fuel and ammunition, is seen to have the greatest sustainment demand for POL and ammunition.

We aggregated the consumption rates shown in Figure D.1 in Figure D.2 for more direct comparison of ground combat elements' sustainment requirements. The consumption rate of the SBE in heavy combat is seen to be more than twice that of any other Army or Marine Corps ground element in heavy combat. Second highest is the consumption rate of the SBE in sustained operations. Finally, Figure D.2 indicates that the SBE in sustained operations is between the SBE in heavy combat and the SBME in heavy combat. This result motivates the repeated finding that results for the SBE in sustained operations were similar to those for the SBME in heavy combat.

Figure D.1
Ground Element Consumption Rates

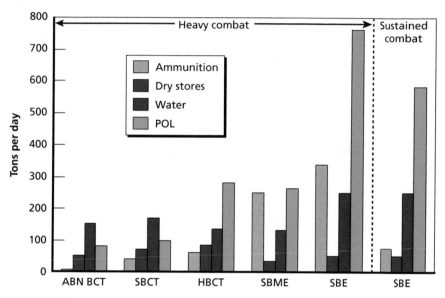

Figure D.2
Aggregate Consumption Rates of Ground Element

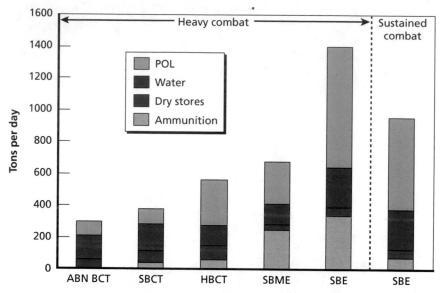

Model Description

Model Overview

The Joint Seabasing Logistics Model (JSLM) was developed by study analysts as a tool to ascertain the feasibility of simultaneously sustaining Marine Corps and Army elements ashore from a sea base or the feasibility of moving an Army element in a reasonable period while sustaining a Marine Corps ground element already ashore. It was designed and developed for this study with knowledge of the uncertainties, unknowns, and analytic simplifications required. The design goal was to provide the ability to examine numerous cases to explore the feasibility of sustainment operations and identify factors critical to their success or failure.

Beyond the ability to address these issues, JSLM was designed to address "what-if" problems and sensitivity issues. What, for example, would happen if water were available to forces ashore? What if a Joint High-Speed Vessel (JHSV) were added as a connector? How would that change sustainability and movement time? What is the sensitivity to Marine Corps and Army elements' distance from the sea base in sustainment? What is the sensitivity to the number of flight spots on the amphibious assault ships for emergency helicopters?

JSLM is an object-oriented model. We can use specific elements of the study to illustrate the object-oriented concept. Amphibious assault ships (LHA(R)s and LHDs) and rotary wing aircraft (CH-53s and MV-22s) are objects in the model acting on each other in well-defined ways; they can be viewed as independent actors with defined roles and responsibilities. They are capable of sending and receiving messages, and of processing data. Similarly, Mobile Landing Platforms (MLPs) and Landing Craft Air Cushion (LCAC) landing craft and Joint High-Speed Vessels (JHSVs) are objects that interact with each other within the model,[1] creating operational friction. As concrete examples, an aircraft returning to the sea base can be advised that there is no landing spot available for it and that it must loiter until one becomes available. Or a fueled aircraft must wait until another has cleared the flight deck before it can take off. Aircraft must conclude operations before the flight window of an amphibious ship ends. A limited number of LCACs can conduct cyclic operations simultaneously on an MLP; others may have to wait their turn. The time for a single LCAC, CH-53, or MV-22 sortie is easy to calculate; it is the friction in the system that drives the problem.

JSLM code is separated into two parts: one that sets up the initial conditions and another representing sea base operations. Within sea base operations, sustainment entails the movement of Petroleum, Oil, and Lubricants (POL); bulk water; ammunition; and dry stores (food, consumables, and spare parts). Each ground element has defined average consumption rates in these categories, according to the nature of the ground element (for example, an Army Stryker Brigade Combat Team (SBCT) or a heavy Brigade Combat Team (HBCT)). Consumption rates depend also on the way in which ground forces are operating: assault operations entail higher consumption rates of fuel and ammunition than do sustained operations ashore. Supplies ashore of POL, water, ammunition, and dry stores are monitored, and additional supplies are brought in to maintain required days of supply (DOS).

Sustainment has first priority in the model; connectors are used only to move Army forces ashore after sustainment requirements are met. In assigning loads, recognition is given to the special capabilities of each connector. For example, JSLM uses LCACs preferentially to transport vehicles ashore; MV-22s are used preferentially to transport troops ashore; and so on.

---

[1] The object-oriented paradigm is supported using the high-level programming language C++, which has features facilitating object-oriented programming.

Operationally, JSLM can be described as a time-stepped deterministic simulation. The *time step* is one minute, which is needed to capture such brief events as the delay experienced by one aircraft ready for takeoff as another aircraft takes off. Because it is *deterministic*, the model cannot directly treat such issues as equipment failure or aircraft losses to antiaircraft systems. The model can be used to treat them indirectly through such techniques as reducing the number of operational LCACs or aircraft at the outset. The model cannot be used directly to treat moving ground forces for which the distance from the sea base is not fixed. Such movement can be treated indirectly by changing distances and rerunning the model. Equally important, there is no provision for treating variation in daily sustainment demand (such as days with abnormally high consumption rates of bulk POL or ammunition).

In cases involving sustainment only, the model is run over a period long enough to average out some modeling artificialities.[2] In cases involving moving the Army through the sea base, the model is run until Army movement is completed.

Model Procedures

As stated above, JSLM is organized explicitly into two procedures: setup and operations, with setup completed before operations begin. This section describes setup and operations procedures within JSLM. It then briefly describes data for the model.

Setup Procedure

Setup consists of preliminary administrative functions; identification of the scenario to be treated with geographic distances; and inputting the parameters for the Marine Corps and Army ground elements, the sea base, and for any JHSVs. It is done entirely from data, rather than by building such functions into the code. Unless stated otherwise, the following definition of characteristics is from data.

---

[2] This practice could be used to accommodate ground force movement.

Administrative functions include assigning a user-selected name to an output file and opening it.

The three analysis cases are indicated to the model. In response, the model prints a message to the output file briefly describing the scenario, to avoid any confusion in interpreting results.

Geographic distances from the sea base to the elements ashore and to a seaport of debarkation (SPOD) are then read from a data file. For flexibility, the model is designed to accommodate multiple SPODs (such as one for Marine Corps elements and a second for Army elements); this capability is not used in this analysis.

Because the Marine Corps element is established on the ground at the outset of the simulation process, its makeup is irrelevant to the problem. It is characterized instead through its daily consumption rates of POL and water (both in gallons per day) and ammunition and dry stores (both in tons per day). In addition, the number of days of supply required by the Marine Corps element is established.[3] The model assumes that the Marines begin with the required DOS for POL, water, ammunition, and dry stores.

Because the Army element may pass through the sea base, its makeup can be highly relevant to the problem. The Army element's makeup is characterized in terms of tons of vehicles, number of personnel, and tons of supplies (dry stores from the perspective of the sea base). The Army element's daily consumption rates and required DOS are characterized exactly as for the Marine Corps element. In addition, as with the Marine's, Army elements begin with the required DOS for POL, water, ammunition, and dry stores.

The portion of the sea base relevant to this analysis consists of Mobile Landing Platforms (MLPs), Landing Craft Air Cushion (LCAC) vehicles, Amphibious Assault Ships (LHA(R)s and LHDs),

---

[3]   In fact, the number of days of supply to be maintained is irrelevant to the problem so long as we assume that the ground elements begin with that supply level. In sustainment, JSLM is essentially a "pipeline" model attempting to flow consumables ashore at least as fast as they are being consumed. DOS levels are included in model output to ease interpretation of results.

CH-53 helicopters, and MV-22 aircraft.[4] Large, Medium-Speed Roll-on, Roll-off (LMSR) and T-AKE ships are represented only implicitly. It is assumed that LMSR ships can transfer vehicles, ammunition, and dry stores onto the MLPs at least as quickly as connectors can take them ashore.[5] Legacy Maritime Prepositioning Force ships are irrelevant to the study; existing pre-positioning ships cannot offload sustainment material at sea and cannot transport Army forces through the sea base.

LCACs and aircraft are tracked individually in the model. Individual LCACs are assigned to the MLPs, and individual available CH-53s and MV-22s are assigned across the amphibious ships.[6] As stated above, there is no consideration for reliability in the model. The technique of reducing the number of LCACs, CH-53s, and MV-22s according to expected availability is used to treat reliability. In addition, some aircraft may be set aside for purposes such as combat search and rescue. The result of all of these offline considerations is number of available LCACs, CH-53s, and MV-22s.

Three MLPs are coded into JSLM. MLPs are characterized as platforms that partially submerge to allow cargo to float on and off them. They would have the ability to "pump up" and "pump down." This is similar to ships such as the 225-meter MV American Cormorant, a float-on/float-off heavy lift semisubmersible vessel, and the pumping characteristics of that vessel were used to represent the MLP. LCACs cannot go aboard an MLP or come off an MLP while it is in the up position. It is assumed that the LCACs carried by the MLPs will need time for maintenance and for crew rest. Accordingly, MLPs are assigned *operating windows* (analogous to flight windows for aircraft carriers or the amphibious assault ships). Finally, MLPs can carry six

---

[4] Aircraft parameters reflecting CH-53K and MV-22 aircraft were provided by MCCDC for this study.

[5] In a sense, this analysis suggests requirements for future Army Prepositioning Ship offload rates. If those future APS ships cannot keep up with connector movement ashore, the APS will become a bottleneck.

[6] The model does not track non-available aircraft such as operationally unavailable aircraft or MV-22s reserved for search and rescue operations.

LCACs but cannot load more than two LCACs at a time. This completes the characterization of the MLP.

LCACs are described in terms of their average speed in operation, their maximum operating hours per day,[7] the number of passengers they can transport in a passengers-only sortie or while carrying vehicles or material, load capacities, and load and offload times for different types of loads. Data used in the model are for LCACs that have gone through the LCAC service life extension program.[8] Finally, what are deemed bias terms are assigned to payload types. These terms elevate the threshold at which LCACs are assigned payload types and thus allow preferences in assigning loads.

Three (LHA(R) and LHD) amphibious assault ships are coded into JSLM; addition of a fourth amphibious assault ship would require code modification. Amphibious assault ships are differentiated in the model by their parameters; no given ship is designated an LHA(R) or an LHD. They are characterized by the number of (port side) spots that can be used for sustainment or to help transport the Army element. They are also assigned a schedule for flight operations, consisting of spotting and unfolding aircraft in preparation for a flight window, the flight window itself, respotting aircraft at the conclusion of a flight window, and a standdown period. This cycle is shown in Figure E.1.

Figure E.1 shows operations over a 28-hour period to clarify the way cycles wrap across days. Again, the figure shows flight operations for representative cycles and aircraft mixes. Times do not need to span entire days; instead, they could be assigned, for example, to limit the flight window to cover of darkness. The number of CH-53 and MV-22 aircraft assigned to each amphibious ship is also an input; the model does not check that they will all fit. Here, each LHA(R)

---

[7] For safety, a limit of 16 hours per day of operation is currently assigned to LCAC crews. LCACs in the model attempt to complete their last sortie of the day within the assigned time limit but are not always able to do so; there can be a "crush" of LCACs at the MLP at times and these crushes are not predictable as LCAC sorties are planned. The operating time limit for LCACs is an input. Entering a value greater than 24 hours can eliminate time limits altogether.

[8] Of note, it is about 10 knots faster than existing LCACs.

Figure E.1
Flight Operations on an Amphibious Assault Ship

RAND *MG649-E.1*

is shown as having been assigned 6 CH-53Ks and 11 MV-22s. The LHD is shown as having been assigned 5 CH-53Ks and 12 MV-22s.[9] Of note, the model "stumbles" somewhat on the first day, because the LHD does not commence flight operations with spotting until 1600 in the model. No sorties are carried over from the previous day. This is one of the artificialities ironed out by running the model for several days.

Once the characteristics of the amphibious assault ships have been established as above, CH-53 and MV-22 aircraft are defined in the model. These are the most complex objects in the models, having complex state vectors and rules governing state transitions and load assignments, and a relatively large number of parameters. The states are respotted and folded, spotting, (re)fueling, shipboard takeoff, conduct-

---

[9] Again, not counting those MV-22 aircraft withheld for purposes other than sustainment.

ing ship-to-ship movement, loading (internally or externally), inbound, landing ashore, offloading, returning to the sea base, loitering (when an aircraft has returned to the sea base but there is no spot open for it to land on), shipboard landing, and respotting.

Rules define the transitions from state to state.[10] For example, from a data file, a representative time for ship-to-ship movement is known. Once the aircraft takes off from an amphibious ship to take on a load from another ship, it is in ship-to-ship movement. A rule uses time of flight to transition the aircraft to loading on the other ship. Flight speeds with internal and external loads are described. Troop, bulk liquid, dry stores, and ammunition load capacities are described. Times to complete the evolutions of refueling, taking off, ship-to-ship movement, loading, landing, and offloading are provided. Finally, bias terms (corresponding to those for LCACs) are assigned to potential aircraft loads. These biases will tend to cause aircraft to be assigned loads for which they are best suited.

JHSVs are characterized through their speed, load capacities, and loading and offloading times. Mixed cargoes such as fuel and ammunition are not allowed. A mean weight is provided for dry stores and vehicle/equipment loads, along with their loading and offloading times. Inputs to the model were taken from a recent RAND Analysis of Alternatives (AoA) for the JHSV. A representative JHSV was used for speed and load representations: an average speed of 41 knots, 15,000 square feet of cargo space, and a maximum troop capacity of 970 passengers.[11] Passenger capacity comes into play when the JHSV has been used to

---

[10] Rules also govern load locations. Aircraft take POL from the amphibious ship, so there is no ship-to-ship movement for POL sorties. All other aircraft loads require ship-to-ship movement. Other than troops, all aircraft loads are external. Aerodynamics limit aircraft speed with external loads (reducing the MV-22's speed advantage over the CH-53K). Rules also apply to internal and external loading, and aircraft operation with internal and external loads.

[11] A draft set of JHSV performance specifications was in circulation at the time of this study. It specifies a threshold troop capacity of 312 passengers. This lesser requirement does not change simulation results, because results did not hinge on the ability to transport all Army troops in just a few JHSV loads.

transport troops to the sea base, so it is assumed that the JI ISV is configured for maximum troop capacity.

This completes model setup prior to the simulation of sea base operations.

Operational Procedures

To address the feasibility of simultaneous sustainment operations, the program runs sustainment operations at the highest possible tempo. If the days of supply for the Marine Corps and the Army elements do not drop below the required DOS, simultaneous sustainment is possible. Otherwise (i.e., operating at a maximum tempo, the required DOS cannot be maintained), simultaneous operation is not sustainable. As stated, the model is written so that any surplus sustainment capacity goes to dry stores (DS; with excess dry cargo building up). This is illustrated in portions of sample model output shown in Figure E.2.

The model output below shows ten days of sustainment for an SBE and an Army airborne brigade in Scenario A, with LCACs used to sustain the MEB.[12] The first day of sustainment is day 0, and end-of-day DOS are shown for each sustainment category for the Marine Corps and Army elements. At the bottom of the figure, following day-by-day material, there is a summary of the number of LCAC, CH-53K, and MV-22 sorties by load category, as well as the levels at which Marine Corps and Army ground elements were sustained.[13]

The high-level performance metric used in the analysis—relative throughput rate—is shown at the bottom of the output.

---

[12] Again, ten days of sustainment operations are used to average out modeling artificialities; they are not intended to represent some operational reality. In addition, it is recognized that the sea base could not provide hundreds of days of dry cargo over this period; the buildup of excess dry cargo quantifies surplus connector lift capacity.

[13] To assist in interpreting output, JSLM outputs material describing the scenario being run. It also outputs inputs, including distances, sustainment rates, and MPG data. For brevity, this output is not shown here. In addition, JSLM creates logs of LCAC, MLP, rotary wing aircraft, and LHA(R)/LHD activities on a minute-by-minute basis. This extensive file is not shown.

Figure E.2
SBME and Army Airborne Brigade in Scenario A

End of day 0
   MEB POL DOS: 10.06; H2O DOS: 10.20; Ammo DOS: 10.24; DS DOS: 47.75
   BCT POL DOS: 5.06; H2O DOS: 5.08; Ammo DOS: 5.08; DS DOS: 9.33
End of day 1
   MEB POL DOS: 10.05; H2O DOS: 10.15; Ammo DOS: 10.24; DS DOS: 94.63
   BCT POL DOS: 5.03; H2O DOS: 5.06; Ammo DOS: 5.16; DS DOS: 14.67
End of day 2
   MEB POL DOS: 10.08; H2O DOS: 10.19; Ammo DOS: 10.24; DS DOS: 141.50
   BCT POL DOS: 5.04; H2O DOS: 5.04; Ammo DOS: 5.24; DS DOS: 19.78
End of day 3
   MEB POL DOS: 10.04; H2O DOS: 10.14; Ammo DOS: 10.23; DS DOS: 188.37
   BCT POL DOS: 5.06; H2O DOS: 5.09; Ammo DOS: 5.32; DS DOS: 25.22
End of day 4
   MEB POL DOS: 10.05; H2O DOS: 10.19; Ammo DOS: 10.23; DS DOS: 235.24
   BCT POL DOS: 5.15; H2O DOS: 5.01; Ammo DOS: 5.40; DS DOS: 30.34
End of day 5
   MEB POL DOS: 10.05; H2O DOS: 10.20; Ammo DOS: 10.23; DS DOS: 281.36
   BCT POL DOS: 5.05; H2O DOS: 5.04; Ammo DOS: 5.48; DS DOS: 36.56
End of day 6
   MEB POL DOS: 10.05; H2O DOS: 10.15; Ammo DOS: 10.23; DS DOS: 328.23
   BCT POL DOS: 5.02; H2O DOS: 5.07; Ammo DOS: 5.56; DS DOS: 41.89
End of day 7
   MEB POL DOS: 10.08; H2O DOS: 10.19; Ammo DOS: 10.23; DS DOS: 375.10
   BCT POL DOS: 5.04; H2O DOS: 5.05; Ammo DOS: 5.64; DS DOS: 47.00
End of day 8
   MEB POL DOS: 10.07; H2O DOS: 10.14; Ammo DOS: 10.23; DS DOS: 421.97
   BCT POL DOS: 5.05; H2O DOS: 5.08; Ammo DOS: 5.72; DS DOS: 52.56
End of day 9
   MEB POL DOS: 10.06; H2O DOS: 10.18; Ammo DOS: 10.23; DS DOS: 468.85
   BCT POL DOS: 5.15; H2O DOS: 5.08; Ammo DOS: 5.79; DS DOS: 57.89

LCAC loads: 677 (4.0 sorties per LCAC per day)
• 1 (0.1%) bulk POL loads
• 2 (0.3%) bulk water loads
• 0 (0.0%) personnel loads
• 51 (7.5%) ammunition loads
• 623 (92.0%) dry stores loads
• 0 (0.0%) vehicle/equipment loads
• Average hours of operation per day: 18.1
CH-53 loads: 481 (3.0 sorties per CH-53 per day)
• 320 (66.5%) bulk POL loads
• 161 (33.5%) bulk water loads
• 0 (0.0%) personnel loads
• 0 (0.0%) ammunition loads
• 0 (0.0%) dry stores loads
• 0 (0.0%) vehicle/equipment loads
MV-22 loads: 846 (2.5 sorties per MV-22 per day)
• 73 (8.6%) bulk POL loads
• 195 (23.0%) bulk water loads
• 0 (0.0%) personnel loads
• 11 (1.3%) ammunition loads
• 567 (67.0%) dry stores loads
• 0 (0.0%) vehicle/equipment loads

Average throughput rate: 2743.5 tons per day
Relative throughput rate: 280.2%
• Average throughput rate for MEB: 2179.7 tons per day
• Relative throughput rate for MEB: 320.5%
• Average throughput rate for BCT: 563.8 tons per day
• Relative throughput rate for BCT: 188.7%

This output illustrates several points made earlier. With a MEB requirement of ten DOS ashore and an Army requirement of five DOS ashore, the sea base was able to sustain the required DOS in POL, water, and ammunition with a significant surplus connector capacity (equivalent to hundreds of days of dry stores built up over a ten-day period).

Looking at the load results, we can see that most LCAC and MV-22 aircraft sorties were to move dry stores (i.e., to represent excess capacity).[14] On the other hand, no CH-53K sorties carried dry stores. With a significant spare capacity seen for transporting dry stores, the CH-53K is clearly more productive than the MV-22 here.

The case shown above, an SBME and an airborne brigade in heavy combat, was run with the MEB 75 NM from the MLPs. Notice that with LCACs sustaining only the MEB, relative capacity for the MEB (more than 300 percent) is significantly higher than that for the BCT (less than 200 percent).

We now move to Scenario B, in which LCACs can sustain both the MEB and the BCT, using the same geometry. Partial results of this scenario are shown in Figure E.3.

Here, the MEB maintains ten days of supplies in all categories, and all unneeded capacity goes to the BCT.

A more stressing sustainment scenario occurs when all Army sustainment is airborne. Output for this case is shown in Figure E.4. The SBE and SPOD are 25 NM from the sea base, and an HBCT is 75 NM from the sea base.

The result here is that the MEB receives 100 percent of its sustainment requirements and the BCT receives 138 percent of its sustainment requirements (overall, the relative capacity is 111 percent). These percentages are viewed as a minimal performance margin.

---

[14] Recall that JSLM is written so that any surplus sustainment capacity goes to dry stores (with excess dry cargo building up). With 92 percent of LCAC and 67 percent of MV-22 sorties assigned dry stores payloads, and with dry stores piling up ashore, there is significant excess capacity.

Figure E.3
SBME and Army Airborne Brigade in Scenario B

End of day 0
    MEB POL DOS: 10.06;   H2O DOS: 10.14;   Ammo DOS: 10.07;   DS DOS: 10.61
    BCT POL DOS: 5.06;   H2O DOS: 5.14;   Ammo DOS: 11.14;   DS DOS: 35.30
End of day 1
    MEB POL DOS: 10.06;   H2O DOS: 10.18;   Ammo DOS: 10.08;   DS DOS: 10.39
    BCT POL DOS: 5.03;   H2O DOS: 5.11;   Ammo DOS: 10.14;   DS DOS: 72.61
End of day 2
    MEB POL DOS: 10.07;   H2O DOS: 10.13;   Ammo DOS: 10.05;   DS DOS: 10.24
    BCT POL DOS: 5.04;   H2O DOS: 5.12;   Ammo DOS: 9.14;   DS DOS: 109.95
End of day 3
    MEB POL DOS: 10.06;   H2O DOS: 10.18;   Ammo DOS: 10.08;   DS DOS: 10.10
    BCT POL DOS: 5.06;   H2O DOS: 5.01;   Ammo DOS: 8.14;   DS DOS: 147.19
End of day 4
    MEB POL DOS: 10.05;   H2O DOS: 10.12;   Ammo DOS: 10.08;   DS DOS: 10.52
    BCT POL DOS: 5.07;   H2O DOS: 5.01;   Ammo DOS: 7.14;   DS DOS: 184.43
End of day 5
    MEB POL DOS: 10.07;   H2O DOS: 10.17;   Ammo DOS: 10.09;   DS DOS: 10.28
    BCT POL DOS: 5.05;   H2O DOS: 5.15;   Ammo DOS: 6.14;   DS DOS: 221.13
End of day 6
    MEB POL DOS: 10.07;   H2O DOS: 10.12;   Ammo DOS: 10.06;   DS DOS: 10.13
    BCT POL DOS: 5.06;   H2O DOS: 5.13;   Ammo DOS: 5.14;   DS DOS: 258.31
End of day 7
    MEB POL DOS: 10.06;   H2O DOS: 10.16;   Ammo DOS: 10.06;   DS DOS: 10.19
    BCT POL DOS: 5.04;   H2O DOS: 5.16;   Ammo DOS: 5.22;   DS DOS: 295.93
End of day 8
    MEB POL DOS: 10.07;   H2O DOS: 10.11;   Ammo DOS: 10.07;   DS DOS: 10.40
    BCT POL DOS: 5.13;   H2O DOS: 5.13;   Ammo DOS: 5.30;   DS DOS: 333.29
End of day 9
    MEB POL DOS: 10.06;   H2O DOS: 10.15;   Ammo DOS: 10.08;   DS DOS: 10.39
    BCT POL DOS: 5.11;   H2O DOS: 5.08;   Ammo DOS: 5.38;   DS DOS: 370.69

LCAC loads: 676 (4.0 sorties per LCAC per day)
- 5 (0.7%) bulk POL loads
- 21 (3.1%) bulk water loads
- 0 (0.0%) personnel loads
- 1 (0.1%) ammunition loads
- 649 (96.0%) dry stores loads
- 0 (0.0%) vehicle/equipment loads
- Average hours of operation per day: 18.1
CH-53 loads: 481 (3.0 sorties per CH-53 per day)
- 311 (64.7%) bulk POL loads
- 154 (32.0%) bulk water loads
- 0 (0.0%) personnel loads
- 3 (0.6%) ammunition loads
- 13 (2.7%) dry stores loads
- 0 (0.0%) vehicle/equipment loads
MV-22 loads: 953 (2.8 sorties per MV-22 per day)
- 67 (7.0%) bulk POL loads
- 63 (6.6%) bulk water loads
- 0 (0.0%) personnel loads
- 265 (27.8%) ammunition loads
- 558 (58.6%) dry stores loads
- 0 (0.0%) vehicle/equipment loads

Average throughput rate: 2828.5 tons per day
Relative throughput rate: 288.9%
- Average throughput rate for MEB: 684.1 tons per day
- Relative throughput rate for MEB: 100.6%
- Average throughput rate for BCT: 2144.4 tons per day
- Relative throughput rate for BCT: 717.6%

RAND *MG649-E.3*

Figure E.4
Army Heavy Brigade from a Short Distance in Scenario A

End of day 0
   MEB POL DOS: 10.05; H2O DOS: 10.20; Ammo DOS: 10.07; DS DOS: 10.11
   BCT POL DOS: 5.04; H2O DOS: 4.98; Ammo DOS: 5.16; DS DOS: 4.92
End of day 1
   MEB POL DOS: 10.06; H2O DOS: 10.10; Ammo DOS: 10.08; DS DOS: 10.23
   BCT POL DOS: 5.03; H2O DOS: 5.02; Ammo DOS: 5.06; DS DOS: 7.67
End of day 2
   MEB POL DOS: 10.05; H2O DOS: 10.17; Ammo DOS: 10.08; DS DOS: 10.15
   BCT POL DOS: 5.03; H2O DOS: 5.03; Ammo DOS: 5.11; DS DOS: 11.09
End of day 3
   MEB POL DOS: 10.05; H2O DOS: 10.17; Ammo DOS: 10.05; DS DOS: 10.08
   BCT POL DOS: 5.07; H2O DOS: 5.00; Ammo DOS: 5.17; DS DOS: 13.59
End of day 4
   MEB POL DOS: 10.05; H2O DOS: 10.07; Ammo DOS: 10.06; DS DOS: 10.19
   BCT POL DOS: 5.00; H2O DOS: 5.01; Ammo DOS: 5.07; DS DOS: 17.37
End of day 5
   MEB POL DOS: 10.06; H2O DOS: 10.08; Ammo DOS: 10.06; DS DOS: 10.12
   BCT POL DOS: 5.00; H2O DOS: 5.02; Ammo DOS: 5.13; DS DOS: 20.65
End of day 6
   MEB POL DOS: 10.06; H2O DOS: 10.20; Ammo DOS: 10.07; DS DOS: 10.23
   BCT POL DOS: 5.00; H2O DOS: 4.97; Ammo DOS: 5.18; DS DOS: 24.13
End of day 7
   MEB POL DOS: 10.05; H2O DOS: 10.15; Ammo DOS: 10.07; DS DOS: 10.16
   BCT POL DOS: 5.04; H2O DOS: 5.01; Ammo DOS: 5.09; DS DOS: 27.02
End of day 8
   MEB POL DOS: 10.05; H2O DOS: 10.16; Ammo DOS: 10.04; DS DOS: 10.09
   BCT POL DOS: 5.04; H2O DOS: 5.01; Ammo DOS: 5.14; DS DOS: 30.39
End of day 9
   MEB POL DOS: 10.05; H2O DOS: 10.11; Ammo DOS: 10.08; DS DOS: 10.20
   BCT POL DOS: 5.04; H2O DOS: 4.99; Ammo DOS: 5.19; DS DOS: 33.56

CH-53 loads: 1171 (7.3 sorties per CH-53 per day)
- 941 (80.4%) bulk POL loads
- 221 (18.9%) bulk water loads
- 0 (0.0%) personnel loads
- 1 (0.1%) ammunition loads
- 8 (0.7%) dry stores loads
- 0 (0.0%) vehicle/equipment loads
MV-22 loads. 1533 (4.5 sorties per MV-22 per day)
- 415 (27.1%) bulk POL loads
- 246 (16.0%) bulk water loads
- 0 (0.0%) personnel loads
- 376 (24.5%) ammunition loads
- 496 (32.4%) dry stores loads
- 0 (0.0%) vehicle/equipment loads

Average throughput rate: 2240.2 tons per day
Relative throughput rate: 111.4%
- Average throughput rate for MEB: 1433.1 tons per day
- Relative throughput rate for MEB: 100.4%
- Average throughput rate for BCT: 807.1 tons per day
- Relative throughput rate for BCT: 138.4%

RAND *MG649-E.4*

Increasing distances by 25 NM (so that the MEB is now 50 NM from the sea base and the HBCT is 100 NM from the sea base) further increases the level of stress in sustainment, as shown in Figure E.5.

Here, we can see that aircraft sortie rate is reduced. The MEB, which has first priority, still receives 100 percent of its sustainment requirements and hovers at ten days of supply in all categories. The BCT receives just 37 percent of its sustainment requirement and quickly exhausts its supplies other than POL. Inability to sustain dry stores and bulk water is reflected in the day-for-day decline in Army dry stores levels over the first days of operation (with 4 DOS at the end of the first day, 3 DOS at the end of the second day, and so on). Further, the sea base is unable to sustain the HBCT with bulk water (as reflected in the slowly declining level of bulk water for the HBCT).

At the outset of this appendix, we stated that JSLM was developed to answer questions of feasibility (just illustrated) and sensitivity (what-if questions). The ability to address what-if questions is illustrated by modifying the above case with the supposition that the ground elements could make or obtain bulk water—over 90,000 gallons per day—so that the sea base would not have to provide bulk water. Partial output, shown in Figure E.6, was generated by simply zeroing SBE and HBCT bulk water consumption rates.[15]

Aircraft-only sustainment is now much closer to success. The SBE is still being sustained at 100 percent of requirements, and the HBCT is receiving 87 percent of its requirements.

Thus far, JSLM operation has been illustrated just for sustainment-only operations. We now illustrate Scenario C, in which Army elements flow through the sea base, using a Marine Corps SBE sustained ashore and an Army SBCT flowing through the sea base. The distance from the sea base to the MEB/SPOD is set to 25 NM here. With the lowest Marine element sustainment burden, the lighter Army

---

[15] The model still initializes the BCT with 5 DOS in all sustainment areas. The result of zeroing HBCT demand for bulk water is then (artificially) fixing BCT bulk water DOS at 5 days.

Figure E.5
Army Heavy Brigade in Scenario A, from a Greater Distance

```
End of day 0
  MEB POL DOS: 10.04;   H2O DOS:  10.03;   Ammo DOS:  10.04;   DS DOS:   10.02
  BCT POL DOS:  4.42;   H2O DOS:   4.00;   Ammo DOS:   4.00;   DS DOS:    4.00
End of day 1
  MEB POL DOS: 10.03;   H2O DOS:  10.05;   Ammo DOS:  10.06;   DS DOS:   10.05
  BCT POL DOS:  4.23;   H2O DOS:   3.00;   Ammo DOS:   3.00;   DS DOS:    3.00
End of day 2
  MEB POL DOS: 10.03;   H2O DOS:  10.09;   Ammo DOS:  10.07;   DS DOS:   10.07
  BCT POL DOS:  4.04;   H2O DOS:   2.00;   Ammo DOS:   2.00;   DS DOS:    2.00
End of day 3
  MEB POL DOS: 10.03;   H2O DOS:  10.05;   Ammo DOS:  10.06;   DS DOS:   10.09
  BCT POL DOS:  3.85;   H2O DOS:   1.00;   Ammo DOS:   1.00;   DS DOS:    1.00
End of day 4
  MEB POL DOS: 10.03;   H2O DOS:  10.04;   Ammo DOS:  10.07;   DS DOS:   10.12
  BCT POL DOS:  3.63;   H2O DOS:   0.00;   Ammo DOS:   0.00;   DS DOS:    0.00
End of day 5
  MEB POL DOS: 10.04;   H2O DOS:  10.05;   Ammo DOS:  10.05;   DS DOS:   10.14
  BCT POL DOS:  3.42;   H2O DOS:   0.00;   Ammo DOS:   0.00;   DS DOS:    0.00
End of day 6
  MEB POL DOS: 10.03;   H2O DOS:  10.05;   Ammo DOS:  10.07;   DS DOS:   10.16
  BCT POL DOS:  3.22;   H2O DOS:   0.00;   Ammo DOS:   0.00;   DS DOS:    0.00
End of day 7
  MEB POL DOS: 10.03;   H2O DOS:  10.10;   Ammo DOS:  10.05;   DS DOS:   10.19
  BCT POL DOS:  3.00;   H2O DOS:   0.00;   Ammo DOS:   0.00;   DS DOS:    0.00
End of day 8
  MEB POL DOS: 10.02;   H2O DOS:  10.05;   Ammo DOS:  10.06;   DS DOS:   10.21
  BCT POL DOS:  2.80;   H2O DOS:   0.00;   Ammo DOS:   0.00;   DS DOS:    0.00
End of day 9
  MEB POL DOS: 10.03;   H2O DOS:  10.04;   Ammo DOS:  10.05;   DS DOS:   10.06
  BCT POL DOS:  2.55;   H2O DOS:   0.00;   Ammo DOS:   0.00;   DS DOS:    0.00
```

CH-53 loads: 886 (5.5 sorties per CH-53 per day)
• 788 (88.9%) bulk POL loads
• 98 (11.1%) bulk water loads
• 0 (0.0%) personnel loads
• 0 (0.0%) ammunition loads
• 0 (0.0%) dry stores loads
• 0 (0.0%) vehicle/equipment loads
MV 22 loads: 1329 (3.9 sorties per MV 22 per day)
• 659 (49.6%) bulk POL loads
• 274 (20.6%) bulk water loads
• 0 (0.0%) personnel loads
• 337 (25.4%) ammunition loads
• 59 (4.4%) dry stores loads
• 0 (0.0%) vehicle/equipment loads

Average throughput rate: 1642.3 tons per day
Relative throughput rate: 81.7%
  • Average throughput rate for MEB: 1429.7 tons per day
  • Relative throughput rate for MEB: 100.1%
  • Average throughput rate for BCT: 212.6 tons per day
  • Relative throughput rate for BCT: 36.5%

RAND MG649-E.5

Figure E.6
Army Heavy Brigade in Scenario A, Self-Sufficient in Bulk Water

End of day 0
  MEB POL DOS: 10.05; H2O DOS: 10.00; Ammo DOS: 10.04; DS DOS: 10.19
  BCT POL DOS: 4.91; H2O DOS: 5.00; Ammo DOS: 4.00; DS DOS: 4.00
End of day 1
  MEB POL DOS: 10.06; H2O DOS: 10.00; Ammo DOS: 10.06; DS DOS: 10.22
  BCT POL DOS: 4.99; H2O DOS: 5.00; Ammo DOS: 3.82; DS DOS: 3.00
End of day 2
  MEB POL DOS: 10.06; H2O DOS: 10.00; Ammo DOS: 10.07; DS DOS: 10.07
  BCT POL DOS: 4.99; H2O DOS: 5.00; Ammo DOS: 4.86; DS DOS: 2.00
End of day 3
  MEB POL DOS: 10.03; H2O DOS: 10.00; Ammo DOS: 10.06; DS DOS: 10.09
  BCT POL DOS: 4.99; H2O DOS: 5.00; Ammo DOS: 5.06; DS DOS: 1.68
End of day 4
  MEB POL DOS: 10.03; H2O DOS: 10.00; Ammo DOS: 10.07; DS DOS: 10.12
  BCT POL DOS: 4.98; H2O DOS: 5.00; Ammo DOS: 5.15; DS DOS: 1.66
End of day 5
  MEB POL DOS: 10.03; H2O DOS: 10.00; Ammo DOS: 10.05; DS DOS: 10.14
  BCT POL DOS: 5.00; H2O DOS: 5.00; Ammo DOS: 5.10; DS DOS: 1.41
End of day 6
  MEB POL DOS: 10.04; H2O DOS: 10.00; Ammo DOS: 10.07; DS DOS: 10.16
  BCT POL DOS: 4.98; H2O DOS: 5.00; Ammo DOS: 5.16; DS DOS: 1.09
End of day 7
  MEB POL DOS: 10.06; H2O DOS: 10.00; Ammo DOS: 10.05; DS DOS: 10.19
  BCT POL DOS: 4.99; H2O DOS: 5.00; Ammo DOS: 5.11; DS DOS: 0.92
End of day 8
  MEB POL DOS: 10.03; H2O DOS: 10.00; Ammo DOS: 10.06; DS DOS: 10.21
  BCT POL DOS: 4.99; H2O DOS: 5.00; Ammo DOS: 5.07; DS DOS: 0.60
End of day 9
  MEB POL DOS: 10.04; H2O DOS: 10.00; Ammo DOS: 10.05; DS DOS: 10.06
  BCT POL DOS: 4.99; H2O DOS: 5.00; Ammo DOS: 5.13; DS DOS: 0.51

CH-53 loads: 855 (5.3 sorties per CH-53 per day)
- 852 (99.6%) bulk POL loads
- 0 (0.0%) bulk water loads
- 0 (0.0%) personnel loads
- 3 (0.4%) ammunition loads
- 0 (0.0%) dry stores loads
- 0 (0.0%) vehicle/equipment loads

MV-22 loads: 1208 (3.6 sorties per MV-22 per day)
- 674 (55.8%) bulk POL loads
- 0 (0.0%) bulk water loads
- 0 (0.0%) personnel loads
- 403 (33.4%) ammunition loads
- 131 (10.8%) dry stores loads
- 0 (0.0%) vehicle/equipment loads

Average throughput rate: 1581.1 tons per day
Relative throughput rate: 96.7%
- Average throughput rate for MEB: 1194.0 tons per day
- Relative throughput rate for MEB: 100.2%
- Average throughput rate for BCT: 387.1 tons per day
- Relative throughput rate for BCT: 87.4%

RAND *MG649-E.6*

force to be moved through the sea base, and the shortest distance considered, this is a best case. Model output is illustrated in Figure E.7.[16]

Here, we can see that all (3,929) Army personnel could be transported on the first day. As observed in the main body of this report, MV-22 aircraft transported a small fraction of all personnel (the large majority of passengers were moved 24 at a time, by LCACs other-wise loaded with vehicles and material). With 214 LCAC loads, the LCACs had the potential to transport over 4,000 passengers. MV-22s could be used tactically ashore without slowing Army movement.

Figure E.7
SBCT Movement Through the Sea Base

```
End of day 0
    MEB POL DOS: 10.05;    H2O DOS: 10.19;    Ammo DOS: 10.08;    DS DOS: 10.11
    Loaded - BCT pers:    3929;    vehicle tons: 4544;    DS tons:   1036
    At SPOD - BCT pers:   3929;    vehicle tons: 4216;    DS tons:   1036
End of day 1
    MEB POL DOS: 10.04;    H2O DOS: 10.14;    Ammo DOS: 10.16;    DS DOS: 10.04
    Loaded - BCT pers:    3929;    vehicle tons: 10816;   DS tons:   1036
    At SPOD - BCT pers:   3929;    vehicle tons: 10460;   DS tons:   1036

BCT movement completed at  2:15:41 (2.65 days)

LCAC loads: 214 (4.7 sorties per LCAC per day)
  • 0 (0.0%) bulk POL loads
  • 2 (0.9%) bulk water loads
  • 0 (0.0%) personnel loads
  • 8 (3.7%) ammunition loads
  • 1 (0.5%) dry stores loads
  • 203 (94.9%) vehicle/equipment loads
  • Average hours of operation per day: 17.1
CH-53 loads: 347 (8.2 sorties per CH-53 per day)
  • 180 (51.9%) bulk POL loads
  • 47 (13.5%) bulk water loads
  • 13 (3.7%) personnel loads
  • 0 (0.0%) ammunition loads
  • 18 (5.2%) dry stores loads
  • 89 (25.6%) vehicle/equipment loads
MV-22 loads: 637 (7.1 sorties per MV-22 per day)
  • 9 (1.4%) bulk POL loads
  • 0 (0.0%) bulk water loads
  • 65 (10.2%) personnel loads
  • 0 (0.0%) ammunition loads
  • 135 (21.2%) dry stores loads
  • 428 (67.2%) vehicle/equipment loads
```

RAND MG649-E.7

---

[16] Output combines MEB sustainment and SBCT movement activities.

Data

This study capitalized upon recent seabasing studies conducted by Marine Corps Combat Development Command (MCCDC) and the Center for Naval Analyses (CNA), and a RAND AoA for JHSVs. The MCCDC and CNA analyses examined sustainment using only aircraft. MCCDC provided the CH-53K and MV-22 parameters used for this analysis. MLP characteristics are based on a *PEO Ships* MLP notional design[17] and the MV *American Cormorant*. LCAC data are from the *MAGTF Planner's Reference Manual*.[18]

MCCDC provided SBE and SBME sustainment requirements for this study. The Army's Combined Arms Support Command provided Army BCT sustainment requirements for this study.

---

[17] "Support Ships," *PEO Ships*, 2007.

[18] MCCDC (revised August 2006c, updated for service life extension).

Bibliography

Clark, ADM Vern, Chief of Naval Operations (CNO), and General Michael Hagee, Commandant of the Marine Corps (CMC), *Naval Operating Concept for Joint Operations*, Department of the Navy, September 2003.

Defense Science Board, Task Force on Mobility, *Enabling Sea Basing Capabilities*, Washington, D.C.: Office of the Under Secretary of Defense for Acquisition, Technology, and Logistics, September 2005.

Department of Defense, *Seabasing Joint Integrating Concept*, Version 1.0, Washington, D.C., August 2005.

Department of the Navy, Office of the Chief of Naval Operations, and Headquarters, U.S. Marine Corps, *Employment of Landing Craft Air Cushion (LCAC)*, NWP 3-02, 12, MCRP 3-31.1A, 1997.

Futcher, LCDR Frank, OPNAV N42, "Seabasing Logistics Concept of Operations," briefing, May 2005.

Hagee, General Michael, CMC, *Concepts and Programs*, Department of the Navy, 2006.

Headquarters, Department of the Army, *Shipboard Operations*, Washington, D.C., Field Manual No. 1-564, June 1997.

Headquarters, United States Marine Corps, *The STOM Concept of Operations (STOM CONOPS)*, Washington, D.C., draft as of April 2003.

———, "Amphibious Requirements: USN and USMC Warfighter Talks," briefing, Washington, D.C., February 2, 2007.

———, *Employment of Landing Craft Air Cushion (LCAC)*, Washington, D.C., MCRP 3-31.1A 1997.

———, *Prepositioning Programs Handbook*, Washington, D.C., PCN 50100234000, March 2005.

Henning, Commander Mark, USN, *U.S. Navy Transformation: Sea Basing as Sea Power 21 Prototype*, U.S. Army War College (USAWC), USAWC Strategy Research Project, March 2005.

Kaskin, Jonathan, OPNAV N42, "The Challenge of Seabasing Logistics," briefing, February 17, 2005.

————, "Seabasing Logistics CONOPs," briefing to NDIA 10th Annual Expeditionary Warfare Conference, October 2004.

Kurinovich, Mary Ann, and Michael W. Smith, *Sustainment of MPF(F) Squadron by T-AKEs*, Alexandria, Va.: Center for Naval Analyses, CAB D0014916.A2/Final, October 2006.

Lambert, Geoffrey C., and Mark M. Huber, "Joint Shipboard Helicopter Operations," *Joint Forces Quarterly*, Winter 2000–2001.

Marine Corps, Programs and Resources Branch, annual updates of MPF(F) program information. As of November 13, 2006: http://hqinet001.hqmc.usmc.mil/p&r/

Marine Corps Combat Development Command—see MCCDC.

McCarthy, VADM Justin, Director, OPNAV N4, "Seabasing Logistics," Presentation to the NDIA 10th Annual Expeditionary Warfare Conference, briefing, October 2005.

MCCDC, *Experimental Marine Expedition Brigade Planner's Reference Guide*, Quantico, Va., 2002.

MCCDC, Mission Area Analysis Branch, "MPF(F) CDD Analysis: Results for Seabasing Capabilities," briefing, March 23, 2006a.

————, "Surface Assault Connector Requirements Analysis Update: Overview to Inform Seabasing Capabilities Study," briefing, April 13, 2006b.

MCCDC, MSTP Center, *MAGTF Planner's Reference Manual*, Quantico, Va., MSTP Pamphlet 5-0.3, 2006c.

National Research Council, Naval Studies Board, *Sea Basing: Ensuring Joint Force Access from the Sea*, Washington, D.C.: The National Academies Press, 2005.

Naval Research Advisory Committee, Panel on Sea Basing, *Sea Basing*, Washington, D.C.: Office of the Secretary of the Navy (Research, Development and Acquisition), March 2005.

Naval Sea Systems Command, SEA 05, *Joint High Speed Vessel Performance Specification (Draft)*, Working Paper, April 2007.

O'Rourke, Ronald, *Navy–Marine Corps Amphibious and Maritime Prepositioning Ship Programs: Background and Oversight Issues for Congress*, Washington, D.C.: Congressional Research Service, RL32513, updated July 26, 2006.

Robbins, Darron L., and Michael W. Smith, *Resupplying Forces Ashore Using Sea-based Aircraft*, Alexandria, Va.: Center for Naval Analyses, CAB D0014746.A2/Final, September 2006.

Schank, John F., Irv Blickstein, Mark V. Arena, Robert W. Button, Jessie Riposo, James Dryden, John Birkler, Raj Raman, Aimee Bower, Jerry M. Sollinger, and Gordon T. Lee, *Joint High-Speed Vessel Analysis of Alternatives*, Santa Monica, Calif.: RAND Corporation, 2006 (not available to the general public).

Schulz, William E., and others, "Skin-to-Skin Replenishment," John J. McMullen Associates, Inc., White Paper, no date.

Secretary of the Army, "The Army Modular Force 2004–2020," briefing, no date.

"Support Ships," *PEO Ships*. As of June 20, 2007:
http://peos.crane.navy.mil/

U.S. Navy, Military Sealift Command Ship Inventory. As of June 20, 2007:
http://www.msc.navy.mil/inventory/inventory.asp?var=DryCargoAmmunitionship

U.S. Navy, V-22 program Web site. As of June 20, 2007:
http://www.navy.mil/navydata/fact_display.asp?cid=1200&tid=800&ct=1

Work, Robert, *Thinking About Seabasing: All Ahead, Slow*, Washington, D.C.: Center for Strategic and Budgetary Assessments, 2006.